Exploiting
Cycle Time
in Technology
Management

McGraw-Hill Engineering and Technology Management Series

Michael K. Badawy, Ph.D., Editor in Chief

Exploiting
Cycle Time
in Technology
Management

Gerard H. (Gus) Gaynor

G. H. Gaynor and Associates, Inc.
Minneapolis, Minnesota

McGraw-Hill, Inc.

New York   St. Louis   San Francisco   Auckland   Bogotá
Caracas   Lisbon   London   Madrid   Mexico   Milan
Montreal   New Delhi   Paris   San Juan   São Paulo
Singapore   Sydney   Tokyo   Toronto

Library of Congress Cataloging-in-Publication Data

Gaynor, Gerard H.
    Exploiting cycle time in technology management / Gerard H. "Gus"
Gaynor.
            p.   cm. — (McGraw-Hill engineering and technology management
series)
    Includes bibliographical references and index.
    ISBN 0-07-023474-4
    1. Technology—Management. I. Title. II. Series.
  T49.5.G38 1992
    658.5'03—dc20   92-16380
                                                               CIP

ISBN   0-07-023474-4

*The sponsoring editor for this book was Robert W. Hauserman, the edit-
ing supervisor was Joseph Bertuna, and the production supervisor was
Donald F. Schmidt. It was set in Century Schoolbook by McGraw-
Hill's Professional Book Group composition unit.*

Printed and bound by R. R. Donnelley & Sons Company.

To my wife Shirley and our children
Mary, Ann, Virginia, John,
Margaret, Katy, and Rosie

Contents

Part 4 Case Histories and Corporate Applications

Series Introduction

Technology is a key resource of profound importance for corporate profitability and growth. It also has enormous significance for the well-being of national economies as well as international competitiveness. Effective management of technology links engineering, science, and management disciplines to address the issues involved in the planning, development, and implementation of technological capabilities to shape and accomplish the strategic and operational objectives of an organization.

Management of technology involves the handling of technical activities in a broad spectrum of functional areas including basic research; applied research; development; design; construction, manufacturing, or operations; testing; maintenance; and technology transfer. In this sense, the concept of technology management is quite broad, since it covers not only R&D but also the management of product and process technologies. Viewed from that perspective, the management of technology is actually the practice of integrating technology strategy with business strategy in the company. This integration requires the deliberate coordination of the research, production, and service functions with the marketing, finance, and human resource functions of the firm.

That task calls for new managerial skills, techniques, styles, and ways of thinking. Providing executives, managers, and technical professionals with a systematic source of informaion to enable them to develop their knowledge and skills in managing technology is the challenge undertaken by this book series. The series will embody concise and practical treatments of specific topics within the broad area of engineering and technology management. The primary aim of the series is to provide a set of principles, concepts, tools, and techniques for those who wish to enhance their managerial skills and realize their potentials.

The series will provide readers with the information they must have and the skills they must acquire in order to sharpen their managerial performance and advance their careers. Authors contributing to the series are carefully selected for their expertise and experience. Although the series books will vary in subject matter as well as approach, one major feature will be common to all of them: a blend of practical applications and hands-on techniques supported by sound research and relevant theory.

The target audience for the series is quite broad. It includes engineers, scientists, and other technical professionals making the transition to management; entrepreneurs; technical managers and supervisors; upper-level executives; directors of engineering; people in R&D and other technology-related activities; corporate technical development managers and executives; continuing management education specialists; and students in engineering and technology management programs and related fields.

We hope that this series will become a primary source of information on the management of technology for practitioners, researchers, consultants, and students, and that it will help them become better managers and pursue the most rewarding professional careers.

Dr. Michael K. Badawy
Professor of Management Technology
The R. B. Pamplin College of Business
Virginia Polytechnic Institute and State University
Falls Church, Virginia

Preface

Most organizations find themselves in the position of applying simplistic quick fixes to resolve unacceptable levels of business performance. The quick fixes for today generate a new series of quick fixes for tomorrow, and the cycle continues.

Exploiting Cycle Time in Technology Management grew out of a concern that concurrent engineering, with its many different names and allied programs, was being proposed as the latest panacea for improving time-to-market performance. Shortening the time between development and manufacturing would allow a business to reach new heights of performance. Speed-to-market would allow a business unit to dominate a market sector.

Those and other approaches suffer from a major inherent deficiency: Managers embark on a piecemeal approach to resolve major business issues, rather than recognize the importance of managing the system. Single-issue management preempts an integrative and holistic vision of the organization.

When organizations that at one time outperformed the annual growth rate of the economy now fail to meet those expectations, then something that is taking place internally needs correction. In many organizations, changes in economic growth were not considered to be justification for not meeting annual targets. Outside influences were unacceptable reasons for nonperformance. However, past management practices have segmented most business units into microworlds—each of which focuses on its own vision and fails to see the implications for the business system. Policies and practices, instead of aligning the business functions, work at cross-purposes. The dynamics of the system are lost in the complexity of the details.

This book specifically relates to optimizing cycle time, but it is also about the practice of managing. It relates to optimizing cycle time in the technology-related functions, but it is also about managing the business enterprise. Managing cycle time cannot be treated as a single issue without considering the functional interrelations. *Exploiting Cycle Time in Technology Management* is about time, timing, and cycle time. It is about systems, systems thinking, and new ways of thinking. It is about change and the impact of change on business performance. It is about process. It is about results. It is about optimizing the use of business resources in the context of a specific infrastructure. Optimizing cycle time needs new approaches to eliminating the delays that occur between the *thinking about* and the *doing*.

Changing an organization to think in terms of *systems*, rather than the pieces of the system, requires a revolution in the way the organization conducts its affairs. Most people have been conditioned to think in linear terms, a process that begins in childhood and continues throughout life. A must always come before B; step 2 cannot be started until step 1 is taken; cause creates effect. Systems thinking requires a circular approach. Cause creates an effect that creates another cause, and so on. Systems thinking emphasizes the interrelations and interconnections of all of the pieces. It involves an increased sensitivity and awareness of the whole, within as well as outside the immediate microcosm.

My objective in writing *Exploiting Cycle Time in Technology Management* has been to elaborate on the related critical issues and what managers must do to improve their own performances as well as the performances of their organizations. That objective is related to the business system. There are no prescriptions for optimizing cycle time. The subject is too complex; its parts are too diverse. Applying simplistic solutions to undetermined causes in a universe of different business units cannot provide any guidance. Each organization must approach the questions related to cycle time from its own perspective and within the limitations of its business infrastructure and resources.

This book raises issues to stimulate new thinking rather than to prescribe a definitive methodology. Optimal cycle times result from managing the business and not from executive-level dictums. They require a change in the thinking processes, a change in the mindset, and possibly a metamorphosis. Levels of expectation must be increased, and executives and managers must reconceive their roles from storekeepers to proactive managers who participate in and contribute personally to the success of the organization.

The organizational infrastructure, the resources, and the associated activities are the three components of the cycle time model. The flexibility and the discipline of the infrastructure and the allocation and the management of the resources determine how organizations manage activities in order to exploit the benefits of optimizing cycle time. Managing cycle time involves managing activities, managing the process, and managing results. It is not a new program; it is a result of effective management practice.

The book is divided into four parts. Part 1 raises the critical issues that must be confronted if an organization attempts to introduce the concept of system cycle time management (SCTM). Part 2 draws attention to the different types of cycle time ranging from the macro to the micro. It considers business system cycle time, product cycle time, project cycle time, and a special category, *time-to-decision*. Part 3 considers four major issues in building a foundation on which to implement SCTM. The foundation includes focusing on value-adding activities, identifying and eliminating major barriers, implementing a responsive management system, and considering the potential costs associated with accelerating cycle time. Part 4 presents some case histories related to the many critical issues associated with managing cycle time.

In *Exploiting Cycle Time in Technology Management*, I have tried to distill the critical issues of managing the many diverse and complex aspects of the effort. Forty years of experience in research, development, and manufacturing have provided me with a system- as well as a business-oriented perspective. Experience is a great teacher. That experience was with several diverse organizations. It included a 25-year career at Minnesota Mining and Manufacturing (3M), where taking a risk is not an impediment to professional or managerial success, and 5 years of international consulting on the management of technology and related issues.

Acknowledgments

It is impossible to list the individuals who have affected my thinking about managing as a practice; many of them are probably not even aware of the lessons they taught. Most of the lessons were positive, but some were education in what not to do. One's thoughts and attitudes have many unidentifiable sources. My sincere thanks to the many people who supported me in a lifetime and allowed me to try new and different approaches to improving business performance. My parents, family, friends, business colleagues, and other authors in some way influenced my thinking about the practice of managing. It quickly becomes evident that the sources of inspiration are abundant.

My sincere thanks to Dr. Michael K. Badawy, the editor of the McGraw-Hill series on engineering and technology management, for his interest in promoting the concept, for providing topical guidance and constructive criticism, and for developing a concept into the finished product: *Exploiting Cycle Time in Technology Management*. To Robert W. Hauserman, my McGraw-Hill sponsoring editor, my compliments to him and his staff for their high level of professionalism. Finally, to my wife Shirley, my partner who made the decision with me to pursue this effort, my sincere thanks for her interest, collaboration, and cooperation.

Gerard H. (Gus) Gaynor

Exploiting Cycle Time in Technology Management

Issues in Optimizing Cycle Time

Confronting the major issues related to cycle time management requires new thinking. Conventional approaches have not delivered—and have little prospect of delivering—optimal benefit if their implementation is focused only on concurrent engineering. If time is an important element in determining business performance, then all factors affecting cycle time must be brought into the picture. The current literature treats cycle time as though it were a discrete element or an independent variable. Optimizing cycle time depends upon many interacting variables in the business system. It does not result from reorganization or quick-fix programs. The primary requisite is that cycle time and its implications for business performance be clearly understood.

Chapter 1 presents some of the current perspectives on and approaches to managing cycle time. Current approaches lose sight of the fundamental issues of managing. If treated as just another quick-fix, single-issue program without an understanding of the basic requirements, most of the effort will not deliver the expected results. Managerial competence is an essential element for creating change. Optimizing cycle time at all levels in the organization requires discipline and managing according to the three E's: effectiveness, efficiency, and the economic use of all resources.

There is no simple definition of cycle time; appropriately, the definition depends on the situation. Chapter 2 presents the basics of system cycle time management (SCTM) and develops the underlying principles related to time, timing, and cycle time. Speed itself is not the issue; the right speed is the issue.

Chapter 2 describes the three components of cycle time and their impact on performance. That is followed by a description of the types, classes, and elements of cycle time. The basics require that managers recognize that cycle time is not a thing that can be purchased off the shelf. A three-dimensional model showing the components of cycle time is presented.

The question is not whether it is necessary to optimize cycle time; organizations will not survive otherwise. Before management can approach the subject of shortening cycle time, it must find the origins of lost time. Chapter 3 focuses not only on finding the usual lost time that organizations deal with but enters the realm of the hidden lost time beginning with the CEO and flowing throughout the organization. The direct lost hours represent a minor part of the problem; it is the multiplying effect of those direct lost hours, which accrue geometrically and increase the indirect lost hours, that is the problem. Lost time as a single category is too broad to consider. It must be reduced to a set of subcategories before its impact can be measured. Chapter 3 considers the lost time demanded by the system, the lost time created by management, the lost time originated in the functional groups, the time frittered away by employees at all levels, the direct and indirect influences, and other internal and external related issues.

Perspectives on Managing Cycle Time

To the pioneers in the vanguard of industrialization, *time to accomplishment* was a major driving force for success. In recent years, unfortunately, U.S. industry has ignored the concepts and the importance of managing the time to accomplishment, now referred to as cycle time, time to market, simultaneous or concurrent engineering, enterprisewide development, and concurrent process development and by other designations. Businesses not only neglected to follow management fundamentals but also mismanaged their available resources. They disregarded the necessity for a disciplined business posture and failed to recognize the impact of the changing competitive forces on business performance. They also naively assumed that tomorrow would be just a continuation of today. This chapter considers:

- System cycle time management
- Managing cycle time—current practices
- Limitations of current practices
- Benefits of managing cycle time

System Cycle Time Management

Managing cycle time has played, today plays, and will continue to play a major role in business performance, but an organization manages cycle time only when it manages the total business for improved performance. Optimized cycle time results from practicing management fundamentals, not from instituting short-term, quick-fix, single-issue programs. No research is required to acknowledge or demonstrate the necessity for managing cycle time. It is not a new idea, concept, innovation, or management discovery. Henry Ford, Edwin H. Land, and many other industrial pioneers managed cycle time. It was implicit in their management behavior. As an example, 2 years after Alfred P. Sloan gave Charles F. Kettering the go-ahead on diesel engine development, Kettering achieved a major breakthrough in diesel engine technology that revolutionized the railroad industry.

The simplistic approaches associated with what is generally considered as reducing engineering or development cycle time must be augmented by applying the principles of cycle time management throughout the organization. Cycle time management must involve more than decreasing the time taken to accomplish a specific activity. Directing energy and effort solely to reducing the time will provide neither a competitive advantage nor an adequate return to the stakeholders. Managing cycle time requires a system perspective—a wide-angle view. System cycle time management (SCTM) is not limited to managing any one time period in research, development, manufacturing, or marketing. SCTM requires managing *all* the activities from concept to commercialization, whether related to a product, a process, or a service, at the most effective rate based on the availability of resources. It also involves managing the cycle time of every business activity that involves some form of implementation. It means managing time, timing, and cycle duration as three distinct yet integrated elements throughout the total organization. It requires attention to managing the system and not just the parts. Decreasing the research and development (R&D) time may or may not provide a business advantage. Research and development are not independent functions. They require additional business resources. They require managers to ask these questions: What is the appropriate or optimal time for performance? What is the required business timing? What is the optimal cycle time for making the best use of business resources?

So where does managing the elements of cycle time fit into the business equation? SCTM requires that activities be organized and performed differently. That, in turn, requires a rethinking and a reconceptualization of the three elements for a systems approach to cycle time management. SCTM can be described as:

> The process for managing the three elements of business cycle time: managing time, managing timing, and managing the duration of each cycle. It encompasses the total organization and pervades every activity, within as well as outside of the immediate business unit, that in any way affects performance.

There is no doubt that some benefits accrue from using the principles of simultaneous engineering and its related concepts, but that approach includes only part of the organization. To gain the benefits of SCTM, the total organization must be involved, beginning with the CEO and including all of the participants.

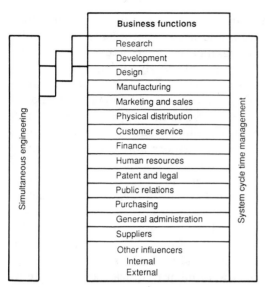

Figure 1.1 Differences in scope between simultaneous engineering approaches and system cycle time management.

Figure 1.1 illustrates the differences in scope between SCTM and the many approaches and variations described as concurrent or simultaneous engineering. Simultaneous engineering is usually limited to considering the relations among research and development, development and design, and design and manufacturing as shown on the left side of Fig. 1.1. System cycle time, as shown on the right side of Fig. 1.1, involves all the activities in all of the business functions. SCTM must be applied to all activities and be practiced by all participants who in any way affect business performance. The concepts must be applied to the business as a system rather than to selected operational functions. Opportunities for improving and subsequently optimizing the three elements of SCTM lie in every personal and group activity and interface. Achieving the total benefit must include more than involvement of the technologically related functions of research, development, and manufacturing.

Single-Issue Management and Functional Focus

As a generalization, executives and managers tend to enshrine the quick-fix gurus. In the process, they lose sight of the fundamentals of both management theory and practice. The basics of management practice are replaced by acceptance of quick fixes that result in a total loss of discipline in the thinking and decision processes of the multi-layered and microsegmented organizational structures. The fundamentals of managing business units effectively, efficiently, and with the economic use of all resources (the three E's)[1] give way to the development of permissive cultures that ignore the importance and necessity of continually improving performance—a condition essential not only for business growth but also for sustaining a viable and functioning organization. In the process, business units, through their executives and managers, direct greater attention to the peripheral, the routine, the administrative, and the justification aspects of business performance, rather than to the creative, the innovative, the proactive, and the leadership activities. The realities and the consequences of their decisions are often disregarded. Self-admiration, self-approbation, and self-justification take precedence over reality, but they provide no guidance for the genesis of a business renaissance.

Business history shows that the number of quick fixes tends to increase geometrically with time. Organizations become infatuated with the latest management guru and conclude that salvation has finally arrived. As an example, *In Search of Excellence,* by Peters and Waterman[2], brought to industry's attention the potential impact of pursuing *excellence* to achieve gains in business performance and productivity. The authors selected eight operational attributes and six measures of long-term financial performance as classification criteria. The study identified 35 companies that passed the test for *excellent* performance in the period 1961 to 1980. The criteria for classification as *excellent* are arbitrary, and one can argue the use of the word "excellent," but all 35 were well-respected organizations. However, excellence is just one of many single issues that business accepted as quick fixes for fundamental and generic problems. Not every action or outcome is entitled to a standing ovation. In an effort to provide positive reinforcement, managers inadvertently praised performance that should have been constructively criticized and corrected.

During the past several years, many of the organizations cited by Peters and Waterman as excellent have been undergoing significant reevaluation and reorganization. Those actions are not just the result of the general decline of the world economy or the increase of global competition. Such organizations as Eastman Kodak[3], IBM[4], Unisys[5], Control Data[6], Digital Equipment[7], AT&T[8], and Motorola[9] have been undergoing various degrees of reorganization that manifests itself in a significant reduction in personnel at all levels. In the international business arena, such European organizations as Compagnie des Machines Bull of France[10], Olivetti of Italy[11], and Philips[12] of The

Netherlands were planning or were in the process of making major staff reductions at the end of 1990. The actions began to take place only after financial reports indicated that profit margins were decreasing. Executives and managers did not anticipate the changing economic environment in spite of all the available information.

Business restructurings, downsizings, early retirement programs, mass terminations, and other cost-cutting programs are a result of many years of misguided short-term and misdirected quick-fix management. The events taking place in many businesses today result from ill-conceived programs that could not pass the simplest tests for financial or social justification. Nonperformance cannot be justified or rationalized for any significant length of time; eventually the facts must be recognized. The quick-fix, short-term approach is not a substitute for managing in accordance with fundamental management principles.

In spite of all the evidence against those approaches, managers in the best of the best business organizations continue to try to resolve major operating difficulties with simplistic techniques or prescriptions. The one-minute management approach does not build a successful organization. Anyone who thinks otherwise need only read the headlines in business periodicals, such as *The Wall Street Journal, Business Week, Forbes,* and *Industry Week,* the academic and professional journals, such as the *California Management Review, Harvard Business Review, Sloan Management Review,* and *The Executive,* and the professional journals of business-related professional societies.

Nevertheless, continued preoccupation with the quick-fix programs absorbs much of management's time and energy. As an example, quality continues to take a high place on the executive altar. It is joined by such icons as strategic planning, information, technology, innovation and entrepreneurship, empowerment, organizational structure, and culture that are supposed guarantees of building business units that will be globally competitive. Academic and professional gurus continue to preach their single-issue gospels as ways to salvation. There is no single and simplistic recipe for either the development or the salvation of a business unit's competitive position. The fact remains that no single approach, if implemented independently, will have a significant positive impact on long-term business performance.

Most managers continue to view their roles as managing microsegments of a much larger entity. They do not think in terms of system needs. But that parochial and segmented view is no longer viable. Realistically, it never was viable except in the rare situations in which businesses controlled both the markets and the resources and neither effectiveness nor efficiency was an important factor. If the bottom-line figures show that the organization has met the forecasts, top management is not concerned about sales that never materialized or cost reductions and efficiencies that could have been instituted. Little if any attention is directed to the *possible*. Actually, that inattention provides the easiest way to climb the corporate ladder and gain recognition and personal satisfaction. Corporate management fosters, encourages, and rewards response to the short term. Focusing attention on single issues in some microsegment of the organization takes precedence over focusing on total business performance. But a competitive position in a global marketplace will not be gained by managing selective issues. No long-term advantage will be gained by managing the business icons as independent, individual, and isolated elements of a business. All must be included in a systems approach to managing and optimizing business performance.

During the 1970s and 1980s, single-issue and functionally oriented management contributed directly to the decline of industrial productivity. The decline started at the top of the organizational pyramid. Managers began working more like hired hands than professionals. They were caught up in the euphoria of their own "yuppiedom" and "charm schools" and "game playing" and disregarded the fundamentals that are essential to meeting business objectives. They expanded the operational comfort zone by functioning as administrators rather than by facing up to the complexities involved in building viable and sustaining businesses. They built bureaucracies that challenge those of the federal government. As executives and managers, they did not provide the necessary leadership. Gaining the benefits of implementing SCTM is not another single-issue management program; it is a result of managing the enterprise as a system—holistically and as an integrated function.

Managing Cycle Time: Current Practices

Although such concepts as simultaneous, concurrent, and parallel engineering have been proposed and reported in the business press, very little effort has been directed by businesses toward taking a broader perspective to managing the multitude of cycle times that occur in any business operation. The use of teams in shortening cycle time also has received considerable attention, but primarily from the viewpoint of technology transfer and not from that of building improved working relationships. Both approaches lack the force for creating major changes in the way organizations manage their operations. Although some benefits may be achieved by implementing limited concepts of managing engineering-related cycle time, much more remains to be gained by pursuing system cycle time management that takes the needs of the total business system into consideration.

There is no doubt that some CEOs are searching for and exploring ways to shorten product development cycles. In a study by United Research Corporation[13], 6 of 10 CEOs listed shorter product development cycles as vital to their future business performance and 7 of 10 expected to reduce the time to market by 10 to 20 percent in the next 2 years. Such statistics may be encouraging, but it is doubtful that targets at that level will enhance business performance or competitive position. Managing cycle time in that context continues to be treated as an independent single issue: Find a way to shorten cycle time.

Eaton[14] discusses the use of simultaneous engineering principles at General Motors in the design of the Chevrolet Corsica and Beretta, in the GM Truck Group, and at GM's Saturn Corporation. He concedes that Opel has used those principles for many years, but in a situation in which product engineering, manufacturing engineering, and design all report to the same individual. Team activities are emphasized, but within the design-to-manufacturing context. The related operations in the concept to commercialization cycle of a new product are not considered. The process emphasizes the engineering-related activities. That approach undoubtedly provides certain operational economies and benefits, but it falls short of achieving the effective and efficient use of resources. Clark and Fujimoto[15], after examining the effects of product content and scope and organizational capability for quick development in the automobile industry, conclude that the lead time advantage of the Japanese is of the order of 12 months.

However, such headlines as "Engineering: Where Competitive Success Begins" and subtitled "The Trick Is to Integrate People, Technology, and Business Systems—And Get to Market First" are not uncommon. In his article, Teresko[16] states that at Hewlett-Packard (HP): "The goal must be to destroy the barriers separating design engineers, production engineers, and manufacturing engineers." The article emphasizes that HP is forging a link between engineering and manufacturing. The objective is that each group loses its independent "look and feel" and functions as a unit that can leverage the use of its resources. Although stressing unity of purpose is essential to all business performance, stating that each group should lose its independent look and feel may be counterproductive. An environment of *sameness* does not release creative minds or foster innovation and entrepreneurship. The same look and feel may just build another immovable and sluggish bureaucracy. HP has also placed a priority on creating a similar linkage with marketing. The emphasis on reducing cycle time essentially focuses on reducing product development cycle time, which is only part of the answer.

Xerox[17] has been trying to speed the transfer of technology from research to commercialization through what it calls an Express Project. Representatives of research, development, manufacturing, marketing, and the customer have input to the system and are involved in every step of the process, each of course, to the extent required at any specific time in the project. To focus attention on the importance of cycle time, Xerox selected a group of research managers, academics, and manufacturing division heads who meet several times per year to discuss laboratory-to-market strategies. The group reviews the successful efforts at other companies and obtains input from its own people regarding the problems associated with cycle time. In the process, Xerox has created cross-functional teams that it hopes will minimize the barriers to technology transfer. Cycle time includes more than the time involved in transferring technology or information from research to commercialization.

Don Smith[18], Ford Motor Company's corporate coordinator for design for assembly and design for manufacturing and a proponent of using multifunctional teams, raises the issues of communication in relation to business unit segmentation. He notes that product and all of the processes associated with it must be dealt with simultaneously. The objective is to make sure that everyone understands the interrelations among the often conflicting tradeoffs that occur in any new product introduction. This approach does not take into account the cycle time management of the many supporting functions.

Kuo and Hsu, in "Update: Simultaneous Engineering Design in Japan"[19], call attention to the integration of product and process design, which includes design for manufacturing, design for assembly, design for reliability, and design for automation. Once again, attention is directed at engineering cycle time. Garrett, in "Eight Steps to Simultaneous Engineering"[20], also directs attention to the engineering-related activities.

Vesey, in "Meet the New Competitors: They Think in Terms of Speed-To-Market"[21], draws attention to time to market. He describes time to market as the time that elapses between product definition and product availability. He excludes the time required to reach an acceptable and agreed-upon product definition. He also says that the new competitors are time-to-market accelerators. The emphasis is on speed: speed in engineering, in production, in sales response, and in customer service. Although his approach also includes speeding up the reaction time in sales and customer response, Vesey places significantly greater emphasis on design engineering and the manufacturing relationship with time to market.

Gomory, in "From the 'Ladder of Science' to the Product

Development Cycle"[22], also directs attention to the product development cycle. He makes an interesting observation: that in many cases there is no introduction of a revolutionary technology. He calls it *cyclic development,* which he describes as "competition among ordinary engineers in bringing established products to market." As he says, "The competition is between your car and my car, not my car and your helicopter."

Bower and Hout[23] look at fast-cycle capability for competitive power. They go beyond the traditional approaches that emphasize simultaneous, concurrent, and parallel engineering and place significant importance on the use of business multifunctional teams that really function as multifunctional teams rather than as simple organizational restructurings. Their findings about how fast-cycle-time companies operate are illustrated in Fig. 1.2.

Bower and Hout also cite some interesting statistics related to Toyota and Detroit as Japanese and U.S. automobile manufacturers. Toyota takes 3 years to develop a new car; Detroit takes 5 years. Toyota needs a cycle time of 2 days through the plant; Detroit needs 5 days. Toyota can schedule a dealer's order in 1 day; Detroit takes 5 days. Toyota turns its inventory over 16 times per year; Detroit is limited to 8. Extreme caution must be taken in drawing specific conclusions from such statistics (the authors may not be comparing similar operating situations), but the differences are sufficiently large to require some direct action. The data at least indicate the order of magnitude of the differences.

Senge, in *The Fifth Discipline*[24], raises some issues related to management teams. He suggests that while they are maintaining the appearance of a cohesive team, the participants actually spend their time fighting for turf and avoiding anything that might detract from their

---

1. Organize into multifunctional and self-managing teams.
2. Manage both the cycle time or individual activities and the cycle time of the whole delivery system.
3. Identify where in the system compressing time will add the greatest value for customers.
4. Use a disciplined approach to schedules.
5. Favor teams over functions and departments.
6. Structure work and measure performance differently.
7. View continuous learning as an essential element.
8. Emphasize informal and ad hoc communication and rapid feedback.
9. Reappraise the fast-cycle capability continually.

Figure 1.2 Characteristics of fast-cycle-time companies as suggested by Bower and Hout.

images or the image of their department. They are supposed to sort out the critical and complex issues of the organization, but in the process they end up with compromises and decisions that all can accept but that have little impact on the organization's performance. Senge quotes Harvard's Chris Argyris that: "Most management teams break down under pressure. The management team may function quite well on routine issues. But when they confront complex issues that may be embarrassing or threatening, teamness seems to go to pot." Argyris' comments were specifically related to management teams, but the same criticism can be applied to most multifunctional teams. Senge suggests that there is a myth to teamwork: "Most teams operate below the level of the lowest IQ in the group." He suggests that the result is skilled incompetence in which people in groups grow incredibly efficient in keeping themselves from learning. Senge goes on to ask the question: How can a team of committed managers with individual IQ's above 120 have a collective IQ of 63?

On two consecutive days, *The Wall Street Journal*[25,26] reported that Digital Equipment found a glitch in its new mainframe computer and would require retrofitting and that IBM had again fallen behind on a key piece of software. At the end of 1990, IBM's software project was already 9 months behind schedule and probably would not be completed until well into 1991. Both companies not only use but stress and publicize their use of multifunctional teams. The costs of the cited delays are undoubtedly high in terms of lost sales revenue, added cost, and impact on customer satisfaction. No specific reasons are provided for the delays, but anyone who has ever managed a major project could easily prepare several scenarios. Simultaneous engineering, with its emphasis on teams, will not eliminate such delays in new-product introductions. The problems begin at the top of the organizational pyramid. There is more to managing cycle time than setting dates for new-product introductions. Both IBM and Digital are committed to using the multifunctional project teams. Both CEOs are committed, but something is missing in the implementation process.

In spite of Senge's criticism of teams, there are situations in which the intelligence of the team exceeds the intelligence of the individuals and the teams have developed extraordinary capacities for coordinated action. Such situations do not occur over night, however. Anyone who has been part of such a team, when coordinated action was directed toward some specific major goal, understands what can be accomplished when the team is focused on meeting the objectives and in the process keeps the total picture in the forefront. All the factors that provide for effective, efficient, and economic use of resources flow out of not only the particular expertise of each individual, but the relations that bind a group as it focuses on accomplishing its goal. All the human charac-teristics, knowledge, experiences, and skills converge on resolving a complex problem or initiating imaginative and innovative approaches for improving business performance. Vitality and excitement exist not only because of the project content but because of the synergy that de-velops among the participants and builds toward a mutual partnership in the truest sense. Not many management teams or multifunctional project teams respond or function in that manner.

Thomas[27] describes what he calls total cycle time (TCT) as the time taken from expression of a customer's need to satisfaction of that need. In the TCT period, he includes all the discrete activities, each with its own cycle time. He emphasizes shortening the individual cycle times, and he suggests that organizations combine a culture-change process with these three distinct cycle time loops: the make/market loop, the design/development loop, and the strategic thrust loop.

- The *make/market loop*—from customer need identification to prod-uct/service delivery—includes the day-to-day activities that are re-quired to deliver the current products and services from the time a need is evident until the revenue is collected.

- The *design/development loop*—from new product concept to cost-ef-fective production—consists of all the activities necessary to develop new products or services from the time a need is evident until the product or service is developed and reaches cost-effective delivery and serviceability.

- The *strategic thrust loop*—from new business identification to busi-ness as usual—encompasses the activities required to start a new business from the time an opportunity arises until the new business achieves viability.

Thomas describes *viability* as the point at which these new businesses develop self-sustaining make/market and design/development loops. He goes beyond the approaches of simultaneous engineering but stops short of considering cycle time in the context of the business unit as a system. He emphasizes shortening rather than optimizing, and he excludes two of the critical elements of managing cycle time: total time and timing. Shortening a cycle for some activity may not necessarily be the answer unless it fits into the total scheme of the plan. In the United States, as an example, managers push to see something physical: Show me some-thing, give me something to look at and touch and feel. Have you ordered any of the tooling? What is the status of the packaging? Such questions dominate the early design stages. In the interim, the time it takes to do the up-front work thoroughly gives way to taking shortcuts that often require development of totally new concepts or the rework of parts or assemblies already released to production. The time from the expression of a customer's need until that need is satisfied represents only a fraction of the time that must be considered in cycle time management. Total time and timing are also part of the cycle time equation.

Limitations of Current Practices

The foregoing commentary and examples demonstrate the limitations of managing engineering-related cycle time only as a selected single issue and an independent approach to improving business performance. Applying the principles of simultaneous engineering may reduce the product development time, but the gains are modest compared with the gains that could be realized if cycle time were viewed from a systems perspective. The limitations occur because simultaneous engineering:

- *Neglects the strategic implications* of managing business system cycle time instead of engineering-related cycle time.

- *Considers managing cycle time* as just another single management issue looked on as a temporary quick fix without any longer-term sustaining power.

- *Attempts to impose a new program* without recognizing the additional educational requirements, attitudinal changes, and work ethic that must be introduced throughout the organization, beginning with the CEO.

- *Excludes many of the key business participants* from the activities and decision streams early in the project. Involvement includes more than awareness or passive association: It requires active participation.

- *Omits consideration of the effects of the policies and practices* of the various functional groups and the significant negative impact of delays in the time to decision by participating specialists and by all levels of management.

- *Disregards many of the limitations and constraints* that both the internal and external system impose on the participants.

- *Assumes levels of economic and social stability and consistency* regarding information related to markets, customer needs, technologies, and other business activities that may or may not be justifiable—especially in long-term projects.

- *Leaves the malpractices of managing a business as a system intact.*

There is little impact on reducing the interfaces and the functional empires that dominate much of U.S. business.

Benefits of Managing Cycle Time

All the references cited so far have identified some benefit from practicing the fundamentals related to simultaneous engineering, but most histories are based on isolated programs within large organizations. They relate to major projects and not to whole organizations. Thomas, in "Executive Weaponry: Short Cycle Times Slay Competitors"[28], relates the story of a leading electronics equipment company in which the manufacturing cycle was reduced from 30 days to 1 day in less than a year. During the same period, however, the company made minimal progress on its 23-day order entry cycle—an example of single-issue management. Subsequently, use of some outside skills allowed this company, in several months, to reduce its order entry cycle to less than 6 days. Although such case histories are interesting, they must be viewed with some caution. Thomas does not inform us what was done to reduce the cycle time from 30 days to 1 day. An order entry time of 6 days seems to be incompatible with a 1-day manufacturing cycle.

Some ranges of improvement for reducing cycle time provided by Thomas are given in Fig. 1.3. Thomas could be accused of exaggeration, and he could also be asked to explain how all the benefits can be attributed to cycle time reduction. Both questions may be academic. The fact remains that significant benefits can be achieved if an organization chooses to manage cycle time holistically and as an integrated process. Thomas' figures could be reduced by 75 percent, but the benefits would exceed the net profit of many organizations.

Integration or desegmentation is a necessary requisite for managing

---

1. Decreases in cycle time from 30 to 70 percent
2. Increases in revenue from 5 to 20 percent
3. Reductions in inventory of 20 to 50 percent
4. Reduction of invisible inventories of 20 to 60 percent
5. Blue-collar productivity increases from 5 to 25 percent
6. White-collar productivity increases from 20 to 100 percent
7. A 10 to 30 percent benefit from depreciation
8. Reduced scrap from 20 to 80 percent
9. Improved lead times of 30 to 70 percent
10. Improved time to market from 20 to 70 percent
11. Improved return on assets by 20 to 100 percent

Figure 1.3 Ranges of improvement in performance by focusing on total cycle time management as suggested by Thomas.

cycle time. Cycle time cannot be optimized in oversegmented and over-specialized business functions, but integration involves more than organizational restructuring. There are several types of integration, and all of them must be considered. Whitson, in "Managerial and Organizational Integration Needs Arising Out of Technical Change and U.K. Commercial Structure. Part I"[29], describes three types of integration:

- *Managerial integration* relates to patterns of recruitment, training, and retraining.

- *Organizational integration* considers how departmental or divisional functions are interlinked.

- *Functional integration* includes both managerial and organizational and raises the issues in regard to the skills and responsibilities demanded of the participants.

Whitson notes that, in an evolving corporate strategic sense, integration requires an intellectual perspective with regard to product and process development. Manufacturing strategy must be linked to corporate strategy. Integration not only links research and development policy with marketing and manufacturing functions but also questions the opportunity costs of increasing skill differentiation and functional specialization. Specialization and professionalism are essential, but the interlinking and cross-functional integration of such skills remain a managerial and organizational challenge.

A more focused perspective by Whitson suggests that integration consists of three dimensions: technological, functional/managerial, and organizational. Organizations must link the managerial and organizational levels, develop unified strategies and understand their interdependency, link the strategic and operational levels, shrink the height of the pyramid, develop cross-functional literacy, and change the current approach from skill fragmentation to multiskilled involvement.

Summary

This chapter describes SCTM and presents some of the limitations of the current approaches. Industry has inadvertently ignored the benefits that can accrue when a systems approach is adopted as the management process. The quick-fix, single-issue management approach, without a fundamental grasp or understanding of the impact on the business, has generated additional unimportant and irrelevant activity. SCTM requires that managers:

- *Expand* the narrow descriptions of cycle time that limit the possibilities of improving business performance for the long term.

- *Appraise* business results and the performance of individual and group activities realistically and discourage use of the "only hear good news" syndrome.

- *Refrain* from relying on the quick-fix, single-issue management gurus. That applies to individual consultants, the authors on the business best-seller list, and the accounting firms that have been transformed into technology and management consultants.

- *Review* the cited references that provide a profile of the current methods for and approaches to managing cycle time. They can provide the reader with additional insight to the limitations of managing what has been described as engineering cycle time.

- *Integrate* cycle time management into the business system. Viewed from a system perspective, it *is* the business.

- *Direct* the organization toward SCTM, which takes into account time, timing, and the total cycle time in every business-related activity, requires a holistic approach, and recognizes the influences of cycle time management as practiced by such administrative functions as financial, human resources, patent and legal, public relations, purchasing, and general administration.

- *Recognize* and communicate to all the participants that cycle time can be managed only by managing the organization effectively, efficiently, and with the economic use of resources (the three E's). The factors that affect time to completion cannot be ignored. SCTM is not a new single-issue program.

References

1. G. H. Gaynor, *Achieving the Competitive Edge through Integrated Technology Management*, McGraw-Hill, New York, 1991, pp. 50–53.
2. T. J. Peters and R. H. Waterman, *In Search of Excellence*, Harper & Row, New York, 1982, pp. 9–25.
3. D. Hoffman, "Kodak Layoff: Efforts to Ease the Pain," *The New York Times*, Oct. 1, 1986.
4. P. B. Carroll, "Hurt by Pricing War, IBM Plans Write-Off and Cut of 10,000 Jobs," *The Wall Street Journal*, Dec. 6, 1989.
5. S. Gross, "Unisys Offering Severance in Roseville," *Star Tribune of Minneapolis*, Sept. 30, 1990.
6. S. Gross, "Price to Quit as CDC Chairman," *Star Tribune of Minneapolis*, April 17, 1990.
7. Associated Press, "Digital Equipment Suffers First Loss, Boston Company's 3,000 Job Cuts Compound Quarterly Drop," *Washington Post*, July 26, 1990.
8. J. J. Keller, "AT&T Plans to Trim Staff by 8,500 in 1990; Cuts for 1989 Total 25,000," *The Wall Street Journal*, Dec. 11, 1989.
9. "Motorola Seeking 2,500 Job Cuts," *Star Tribune of Minneapolis*, Oct. 31, 1989.
10. J. Neher, "Bull to Close Plants, Slash 5000 Jobs," *International Herald Tribune*, Nov. 9, 1990.

11. G. Collins, "Italy's Olivetti to Cut 7,000 More Jobs in New Signal of Weak Computer Sector," *The Wall Street Journal,* Nov. 14, 1990.
12. E. A. Stein, "Philips Mum on Impact of Layoffs on U.S. Operations," *Computer Design,* Nov. 12, 1990.
13. T. S. Perry, "...so what's to be done," *Spectrum, Publication of the Institute of Electrical and Electronic Engineers,* October 1990, special issue, p. 61.
14. R. J. Eaton, "Product Planning in a Rapidly Changing World," *Int. J. of Technology Management,* vol. 2, no. 2, 1987, pp. 183–189.
15. K. B. Clark and T. Fujimoto, "Lead Time in Product Development Explaining the Japanese Advantage," *J. of Engineering and Technology Management,* vol. 6, no. 1, 1989, pp. 25–58.
16. J. Teresko, "Engineering: Where Competitive Success Begins," *Industry Week,* Nov. 19, 1990, pp. 30–38.
17. T. S. Perry, "...so what's to be done," *Spectrum, Publication of the Institute of Electrical and Electronic Engineers,* October 1990, special issue, pp. 61–63.
18. T. R. Welter, "How to Build and Operate a Product Design Team," *Industry Week,* April 16, 1990, pp. 35–58.
19. W. Kuo and J. P. Hsu, "Update: Simultaneous Engineering in Japan," *Industrial Engineering,* October 1990, pp. 23–26.
20. R. W. Garrett, "Eight Steps to Simultaneous Engineering," *Manufacturing Engineering,* November 1990, pp. 41–47.
21. J. T. Vesey, "Meet the New Competitors: They Think in Terms of Speed-to-Market," *Industrial Engineering,* December 1990, pp. 20–26.
22. R. E. Gomory, "From the 'Ladder of Science' to the Product Development Cycle," *Harvard Business Review,* November-December 1989, pp. 99–105.
23. J. L. Bower and T. M. Hout, "Fast-Cycle Capability for Competitive Power," *Harvard Business Review,* November-December 1988, pp.110–118.
24. P. M. Senge, *The Fifth Discipline,* Doubleday/Currency, New York, 1990, pp. 1–16.
25. J. R. Wilke, "Digital Finds Glitch in New Computer, Prompting it to Retrofit Every Machine," *The Wall Street Journal,* Dec. 18, 1990.
26. P. B. Carroll, "IBM Says It Has Fallen Behind Again on a Key Piece of Office Software," *The Wall Street Journal,* Dec. 19, 1990.
27. P. R. Thomas, *Competitiveness Through Total Cycle Time,* McGraw Hill, New York, 1990, pp. 8–37.
28. P. R. Thomas, "Executive Weaponry: Short Cycle Times Slay Competitors," *Electronic Business,* Mar. 6, 1989, pp. 116–120.
29. T. G. Whitson, "Managerial and Organizational Integration Needs Arising Out of Technical Change and U.K. Commercial Structures," *Technovation,* vol. 9, 1989, pp. 577–605.

2

Understanding
the Fundamentals

Chapter 1 provided the current perspective on managing cycle time and described the limitations of focusing attention only on the elements related to engineering cycle time. System cycle time management (SCTM) introduces an integrative and holistic approach: integrative among the many microsegments of the organization and holistic from the viewpoint of involving the total organization. SCTM draws attention to the total system rather than only some specific segment or organizational function. By concentrating on the system, opportunities are provided; they allow leveraging not only the use of the available resources but also the use of the organizational infrastructure. Stated in another way, SCTM concentrates effort and energy on making the total greater than the sum of the parts: The objective requires making 2+2 something greater than 4. Chapter 2 lays the foundations for SCTM by considering the following issues:

- Relations of strategy, technology, and cycle time
- The tripartite organizational model
- Three components of cycle time
- Elements of the cycle
- System cycle time model
- SCTM—strategic or operational issue
- Shortening or optimizing

Relations of Strategy, Technology, and Cycle Time

The performance of any business unit ultimately depends on strategy, technology, and approach to cycle time management, but the scope of cycle time must be expanded. Figure 2.1 shows the linkage of the busi-

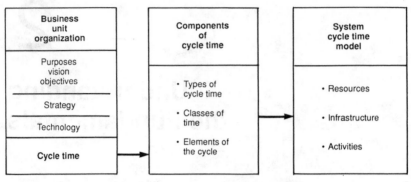

Figure 2.1 Relations of business unit organization, the components of cycle time, and the system cycle time management model.

ness unit organization to the components of cycle time and the SCTM model. Strategy, technology, and cycle time are three common critical issues that determine 95 percent or more of any organization's ability to compete successfully in the marketplace. Strategy is an absolute essential, and as described here it is preceded by and includes the corporate purposes, the vision, and the objectives. They must be clearly delineated, or they will provide no business direction. Wishful thinking on the part of management produces about the same results as the wishful thinking of the gambler, who is confident that the next card or the next roll of the dice will erase all of the previous losses.

The description of technology must go beyond the usual very narrow association of technology with science and engineering or with research, development, and manufacturing. If technology is described as comprising the tools and techniques for accomplishing objectives, then all the business functional groups involve their own specific complement of technologies. Strategy, technology, and cycle time are three interlinked issues in business performance. They cannot be separated and treated as single-issue quick fixes.

Cycle time must take on an expanded meaning and include more than the narrowly described approaches referred to as some form of engineering cycle time. Cycle time comprises three specific components as shown in Fig. 2.1: types of cycle time, classes of time, and elements of the cycle. Each of the three components plays a specific role in determining business unit performance.

System cycle time relates the business infrastructure, the available resources, and the related activities in such a way as to take advantage of the holistic and integrative aspects of the three components of cycle time as shown in Fig. 2.1. Organizational syntax or morphology, the three components of cycle time, and the three-dimensional-system cycle time model interact in the management process.

Tripartite Organizational Model

SCTM will be presented in the context of a generic organizational model: the tripartite organization [1]. The purpose is to expand the understanding, scope, and significance of system cycle time management as it relates to the business system rather than to some limited functional component. The generic model of the tripartite organization [2], Fig. 2.2, embodies all the elements essential for developing a broad perspective of system cycle time management. Engineering cycle time management, as discussed in Chap. 1, is generally limited to the product genesis entity with occasional participation of marketing and sales. To avoid esoteric explanations of SCTM, the descriptions, examples, and commentary will be directed at the organizational structure illustrated in Fig. 2.2. This generic model describes any organizational structure, whether product- or service-oriented. For service-oriented businesses, it requires only changing the description of the manufacturing function. The model recognizes three distinct generic entities that exist in any business organizational structure: product genesis, distribution, and administration.

- *Product Genesis* includes research, development, and manufacturing.

- *Distribution* includes marketing and sales, physical distribution, and customer service.

- *Administration* includes financial, human resources, patent and legal, public relations, purchasing, and general administration.

The three entities shown in Fig. 2.2 are surrounded by strategy, technology, and cycle time, since each impacts each and every function and

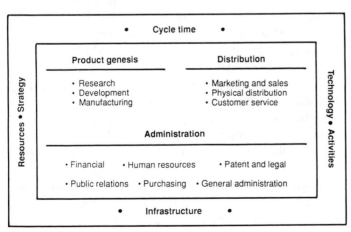

Figure 2.2 Tripartite organization.

activity in the three entities. The three also interact with the three axes of the SCTM model: infrastructure, resources, and activities. A major project within any functional group or at any management level, such as a task force and other special investigatory group, can also be considered as an activity and subdivided as required. It is not important whether a project is a multimillion dollar or multiyear activity or is subdivided into many subprojects of more limited scope.

Strategy involves the total organization and not just some designated executives. An organization needs not only business strategy but also the strategies related to technology and cycle time. A business strategy that excludes technology and cycle time strategy minimizes the impact of two major contributing operational factors. A technology strategy without an interlinked cycle time strategy prevents timely implementation of technology. A strategy that emphasizes fast cycle time to the exclusion of technology provides at best a very short-term benefit. However, strategy, technology, and cycle time must include not only the product genesis entity but also the distribution and administration entities. Distribution and administration have their own unique strategies, technologies, and cycle times [2]. Management sees only part of the picture when it limits the discussion of strategy, technology, and cycle time to product genesis.

The tripartite organization model applies to both product and service organizations.

- A *software development* business, although a service organization, requires the three entities. It has a manufacturing function.

- A *fast-food establishment* requires all three entities. Manufacturing takes place in the kitchen.

- *Service organizations,* such as certified public accountants, attorneys, large consulting organizations, hospitals, educational institutions, and government agencies, fit into the generic organization. Their manufacturing output is information, cured patients, intellectually curious students, and so on.

- *Banks and other financial institutions* also require a manufacturing function. It includes all the activities that are required to move the millions of individual pieces of paper within the organization as well as to the customers. All the manufacturing-related activities apply: planning and scheduling, quality, process, machine and facilities maintenance, and so on.

- *Internal staff functions* also conform to the tripartite service model. Such internal staff services provide added value if they are organized to add value through active participation and acceptance of responsibility for results. If such groups function as active participants rather than as consultants without responsibility for their actions, they add value through their participation. Their production is a measurable output for their input.

Each of the preceding examples involves elements of manufacturing. Whether the output results in a report, an operation in a hospital, a lecture to students, or the development of the infrastructure, all relate to some form of production activity. Something is expected as a result of a predetermined effort. Sometimes the activity manifests itself in a physical product or service and at other times in the initiation, development, or transmission of some idea or concept. Using a little imagination and eliminating the occasional pejorative reference associated with such words as "production" and "manufacturing," as applied to professionals, allows use of the model as a generic one. The tripartite organizational model will be cited throughout this book as a point of reference.

The Three Components of Cycle Time

It is insufficient to discuss cycle time simply as a theoretical concept or construct. Executives and managers, professionals and semiprofessionals in all disciplines, and all the participants must understand the three major components of cycle time as shown in Fig. 2.1. They include:

- Types of cycle time
- Classes of time
- Elements of the cycle

Knowledge of and understanding the three components and their interactions, as well as their role in managing cycle time from a systems perspective, provide specific opportunities for optimizing the resources and taking full advantage of the infrastructure.

Types of cycle time

Businesses are involved in managing different types of cycle time, which can be classified as:

- System
- Product or service or process
- Project

System cycle time. System cycle time relates to the length of time it takes to push something through the system. That *something* could be a new product design, a customer's order, an internal report or study, an engineering change order, a major reorganization, an approval of an investment, or a response to some business-associated tragedy. System cycle time depends on the effective use of policies and procedures, the efficiency of the administrative bureaucracy, the executive and management attitudes, the business operating philosophies, the involvement by and the jurisdiction of staff groups of all types, and, of course, the *creators of delay* who have nothing better with which to occupy their time.

Product, service, or process cycle time.   Product or service and process, as a type of cycle time, considers the elements of time involved in the concept of commercialization cycle. Time begins with the enunciation of the concept and ends with successful implementation or commercialization. Product and service cycles, because of their magnitude, are generally divided into many smaller and discrete cycles associated with the participating functional groups. It must be clearly understood that process cycle time affects product and service cycle time and must be considered as an integral part and not an independent type. Product or service and the associated process cycle time are linked and must be treated as one.

Project cycle time.   Project cycle time refers to the duration of time that is required to move an activity from point *A,* the beginning, to point *B,* the conclusion. The cycle time can be long or short depending on the scope of the defined activity. It could include a multiyear effort to develop a new technology, construct a new factory, introduce a new line of products, make the change to the computer-aided design system, totally restructure the information management system, modify the employee appraisal process, replace a control system in a process, introduce some minor components in a specific product, or initiate a new purchase order system. It includes the activities of all the organizational groups and subgroups.

Classes of time

Time must also be divided by its specific role in managing cycle time. Taking the macro approach and lumping all the different types of time into one category called *time* is insufficient. The classes of time include:

- Predecision time
- Decision time
- Time from decision to activity
- Activity time
- Time from activity to implementation or commercialization

Predecision time. Predecision time includes the time involved in bringing an idea or concept to the point at which sufficient information is available on which to make the appropriate go or no-go decision. It involves an analysis of the risk levels that an organization can accept without a major negative impact on the business in the event of failure.

Decision time. Decision time is the time taken to make the decision. It usually depends on the number of individuals in the decision stream, participant familiarity with the issue, the level of financial support, and so on. The time for decision can be significantly extended if managers are limited in their authority to commit the business unit to a financial expenditure. When a multibillion dollar organization requires expenditures of modest amounts to be approved by a senior management committee or board of directors, the time to decision can be significantly extended. When that same organization requires 16 or more signatures prior to the time when the board is expected to act, additional time delays occur.

Time from decision to activity. The time taken from decision to activity can also create additional delays. If the people resources or the facilities and equipment are not available, delays will occur. When a project spans nations or continents, the time involved in orienting and focusing a group that has never functioned as a partnership may be considerable, to say nothing of the problems associated with dealing with different cultures and the parochialism that exists in most functionally oriented businesses. The urge to reconsider directions after the go-ahead has been given, also can consume valuable time. That does not suggest that new information should not be investigated, but if major changes are required, it may be worth the time and effort to stop the activity and reassess it in its entirety. Although such an approach may create some furor at the executive levels, it may be more prudent to face the fury now instead of when the project is two-thirds of the way to completion.

Activity time. Activity time, the actual time duration of the work effort, has the potential for being affected by many unknowns. Business conditions, loss of personnel, financial setbacks, and miscalculation of levels of technological understanding can affect cycle time. The future cannot be predicted with high levels of confidence. Anticipation of potential difficulties represents the best possible safeguard against such actions. Activity time often includes thousands of distinct activities and microdecisions that can affect performance. Cumulatively these somewhat minor single activities and their related microdecisions can have catastrophic effects on project success. Although any business eventually must be divided into its smaller operational groups, the work efforts in those groups determines the many cycle times of the whole organization.

Time from activity to implementation or commercialization. The continuum from predecision time to implementation or commercialization often results in a stalemate at the end of the activity cycle. That is especially true when a major objective is subdivided into many smaller segments. The elapsed time from completing a multitude of single activities to the integration of their results into some workable configuration often results in significant extensions in cycle time. It is during those periods that the available time of supporting groups must be managed. Organizations do not have the luxury of allowing professionals to sit on the sidelines and wait. Delays occur in all stages of the system, product, and project cycle, and managers must provide value-adding work during those periods.

The five classes of time are introduced not to complicate the considerations of time and its role in cycle time, but instead to focus attention on the need to recognize where time delays originate. Time is part of cycle time management. Cycle time cannot be managed if the time to accomplish certain activities is not factored into the SCTM equation. Managing cycle time is not accomplished by merely accelerating the pace of work or changing the schedule. Time and effort must be spent in determining where the time lags are being created. That can be accomplished by dissecting time into its various classes. Generally, executives direct their total attention to reducing activity time in research, development, and manufacturing from a macro perspective rather than determining where the lost time actually originates. Working from the macro perspectives provides little if any information about how cycle times will be optimized, how timing of activities will be improved, or how the available time will be directed toward value-adding activities.

Elements of the Cycle

The effectiveness of system cycle time management depends on managing three distinct but interacting elements:

- Time
- Timing
- Cycle duration

SCTM requires mastery of those three elements, which work inter-

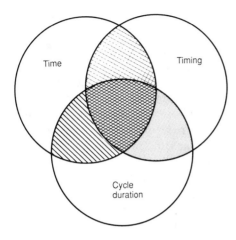

Figure 2.3 The interlocking relations of time, timing, and cycle duration.

actively, collectively, and concurrently as shown in Fig. 2.3. They are not independent: Effective implementation of SCTM requires taking advantage of the interlocking benefits of time, timing, and cycle time duration. Focusing attention on time alone, without considering timing and cycle time duration, may provide more efficiency in manufacturing, but it will have little impact—and certainly no major impact—on total business performance. Emphasizing timing only and disregarding the effects of time and cycle time duration often results in increases in time, that is, increases in the total number of hours for accomplishment. Stressing the importance of cycle time duration to the exclusion of time and timing also results in something less than acceptable performance. The cycle time as described may have been achieved, but if an excessive number of hours are required in that cycle and if the timing is either late or early, the additional costs that will be involved may result in the failure of the endeavor. Time, timing, and cycle time duration interact in SCTM, and each must be appropriately emphasized depending upon the specific peculiarities of the work effort.

The elements of the cycle extend beyond the usual issues associated with managing time, managing management time, time-and-motion study, and so forth. The three elements do not refer only to the CEO, the executives, and the managers of the business unit. The total organization must be involved and must recognize the importance of and the impact of the three cycle elements on business performance. At the same time, understanding the significance and the full meaning of the three E's, *effectiveness, efficiency, and economic use of resources,* that were mentioned in Chap. 1, must be reintroduced into the management vocabulary and be reinforced in all business operations.

Time

Without belaboring the necessity for managing time, certain factors must be considered. Time is a commodity and a business resource, but it must be viewed in perspective. It goes beyond developing time management skills. It cannot be reclaimed or recycled. A great number of books have been written about time and how to make more of it available for constructive work. Recipes and prescriptions are provided and exhortations are issued as to the role that time plays on performance, but the response SCTM cannot be directed at packing more work hours into a day. Thinking about strategy or technology or cycle time while commuting may or may not be productive use of time. Time spent listening to the latest motivational guru while commuting or walking to the cafeteria may be spent more productively by listening to some thought-provoking literature. Speed may be a major factor in repetitive work, but when applied to professional and semiprofessional activities, in which thinking and innovation determine levels of success, slower may be faster. Speed with substance and business discipline provides an advantage, but speed alone results in wasted energy. Our penchant for action rather than a penchant for thought prior to action often creates various types of dysfunctional consequences that only multiply the required time for accomplishing an activity.

Such books as *Right on Time!,* by Lester Bittel [3], relate the traditional approaches to managing time. They are generally prescriptive and deal only with the lower levels of time management. They assume that the system is functioning effectively and efficiently, and the objective is to try to increase the output. Those are essential first steps, but to-do lists are not going to have a major impact on business performance. Organizations must think more comprehensively about the fundamental issues that consume time without adding value. Regardless of how effectively or efficiently a particular activity is executed, it is wasted effort if it adds no direct or indirect value.

The study of time goes beyond the simplistic approaches associated with most business or how-to books. Time has been studied in many of the academic disciplines: cultural anthropology, economics, literature, philosophy, management, sociology, theology, science, engineering, and mathematics. It has received multidisciplinary attention, although not interdisciplinary scholarship. Jean Claude G. Usunier, in "Business Time Perceptions and National Cultures: A Comparative Survey" [4], suggests that time is strongly influenced by cultural patterns. His study in Brazil, France, Mauritania, South Korea, and West Germany shows that developing countries tend to favor ideal economic time, which contradicts their behavior patterns. Although Usunier studied the time perceptions in different national cultures, personal experience shows that the same behavior patterns apply to business organizations. Different functional and subfunctional groups of a business will perceive *time* differently, based upon the subculture in which they are working. Their verbalization of economic time contradicts their behavior patterns. Research is not as time-oriented as manufacturing—a sense of urgency seldom exists in research. Management may be time-oriented in bringing a new product to market, but the participants may not share the same urgency. Time will be perceived differently depending on the subculture developed by the leader. The differences between the actual behavior and the stated ideal must be recognized and resolved.

Hall, in *The Dance of Life* [5], which also considers time from a cultural perspective, makes a distinction between:

- Monochronic time—M-time
- Polychronic time—P-time

M-time appears to be an obsession in western cultures. Such words as scheduling, prioritizing, allocating, compartmentalizing, and classifying direct the activities of M-time cultures. They are linear: A comes before B; time is a resource that is wasted, used effectively and efficiently, or saved. M-time is oriented toward activities, schedules, and adherence to stated procedures. Depending on how it is perceived, M-time can enhance or hinder the performance of an organization.

P-time stresses the involvement and participation of people with great degrees of flexibility regarding time. Prearranged schedules can easily be modified; time is not considered as being wasted regardless of the activity or inactivity; and several activities may be occurring at the same time.

M-time and P-time can be considered as parallels to McGregor's [6] Theory X and Theory Y. They are at opposite ends of the spectrum. Each serves a specific purpose in a specific situation. Neither, by itself, is better or worse than the other. There are many variations of each and, as with all theories or hypotheses, application determines the ultimate value to the organization. However, the main distinction is that polychronic time places a greater emphasis on people and the system. In considering time as one of the three elements of the cycle, it is important to focus attention on P-time. Managing P-time provides greater benefits, because P-time focuses on the individual and the system. It goes beyond the tools and the mechanics of implementing scheduling systems. It goes beyond the mundane exercises in saving time. P-time as used in the context of cycle time considers what, why and why not, how, who, when, and where.

In *Managing Management Time,* Oncken [7] describes approaches that direct the manager's attention to achieving more visible and significant results rather than doing more work in less time. He falls somewhere on the continuum from M-time to P-time as described by Hall. He draws our attention to the dilemma that managers face: dealing with the boss-imposed time and system-imposed time, which are time-consuming and often nonproductive activities that do not add value to the organization. He emphasizes the concept of self-imposed time as a solution to the dilemma. The ability to apply the concept depends on certain personal qualities such as initiative, drive, enthusiasm, loyalty, resourcefulness, guts, imagination, and foresight. The value of self-imposed time generated by those attributes depends on other personal qualities: character, personality, and competence.

Oncken omits one very important attribute that applies to everyone in the organization regardless of position in the hierarchy: the ability to discriminate between what is system- and what is boss-imposed time is important. Early in my career, I was approached by my superior about the fact that I did not submit certain information that he had requested. I responded by asking him whether lack of any of that information was preventing him from making a decision, putting him in an untenable position with others, or delaying any activity. His response was nonverbal: Body language replaced the usual vocal exhortation. He looked perplexed, as if to say, "What right do you have to make that decision?" After awhile he responded only by saying, "I never thought about it that way." A short discussion followed over a cup of coffee, during which we discussed the need for managers and others to be able to gain a sense of the important. He recognized that he was passing on many requests from his own office that were not adding value. As time went on, he also began to question requests for information from above as well as from all requestors.

Timing

No one will question the necessity or the importance of timing on business performance. Timing requires consideration and a positive response from all elements of the tripartite organization. It is not just a matter for consideration by marketing and manufacturing. Emphasis on proper and realistic timing begins at the top of the organization. Timing is one aspect of planning, but a degree of flexibility is required. Managers and individuals at all levels of the organization often complain about the changes that occur after some plans have been approved. Now, there is no set rule for how often plans may have to be modified for many different reasons. Plans that project activities for 6 months or more in any endeavor are bound to change, if for no other reason than that new information is uncovered in the process. The degree to which the plans will change does make a significant difference though, and it depends on the realism and effort that was put forth in developing the original plan and the confidence level that was associated with it. The number of changes generally can be traced to the quality of the up-front work, which included not only a great deal of analysis from all of the participating functions but also the ability of some person or group of persons to synthesize the relevant information into a doable project.

In most situations, timing is considered from the perspective of the manufacturing and the marketing and sales organizations. There is no doubt about the relevance and importance of timing in those two functions. If manufacturing cannot provide a workable and troublefree product on the scheduled date, marketing and sales undoubtedly cannot meet their objectives. Two cases were cited in Chap. 1: IBM's delay in producing software and Digital Equipment's retrofitting its new mainframe computer. If marketing and sales miscalculates the timing for a new product entry, either too early or too late, then valuable resources have been mismanaged. If the product appears too early, it may have siphoned off resources from other products that could have been introduced in a more timely manner. If the product was introduced late, that is, after the competition was able to score an initial foothold by an earlier introduction, then the company will play the catch-up game, which is always costly and time-consuming. The same scenario applies to internal timing of all activities regardless of the source.

Timing of information plays a major role in business performance. As an example, Beeby, in "How to Crunch a Bunch of Figures" [8], discusses the need that business managers have for accurate and comprehensive information. He discusses Frito-Lay's old decision-making structure whereby information crept upward through the organization. The process did not provide timely feedback to management from the field operations. Frito-Lay installed a decision support system (DSS) that now gives 200 managers detailed sales and inventory information fed into the system by 10,000 route salespeople equipped with handheld computers. Beeby cites a situation in which Frito-Lay's DSS took only a couple of weeks to discover a drop in market share of a particular product. Prior to DSS, it may have taken as long as 3 months. The reason was traced to the introduction by a supermarket chain of a store brand product.

There was an interesting response by Chambers [9] to Beeby's article in *The Wall Street Journal* titled "Don't Let Computers Do All Your Thinking." Chambers, a visiting assistant professor at Emory Business School in Atlanta, recognizes the substantial benefits of a well-designed decision support system. He also raises an important issue of becoming overly dependent on such technology as DSS when other solutions may be more appropriate. Beeby, in his article, stated that Frito-Lay's DSS took a couple of weeks to discover a drop in market share. Chambers was surprised that the Frito-Lay contact for that chain did not inform someone the moment the store brand product hit the shelf. In that sense, reliance on the DSS delayed the information. Chambers concluded by stating that: "We must be careful to let the computer replace only the information tools it can outperform. It should never replace our own experiences and judgment. The computer should be only one part of a company's DSS."

One of the benefits of the Frito-Lay DSS as described by Beeby is that it "assists in management by walking around." Such a statement also calls for a response. Beeby, the president and CEO of Frito-Lay, a division of PepsiCo Inc., states: "When Tom Peters [10] coined that phrase, he wasn't thinking of a computer tour of operations by the CEO. But that is what DSS allows me and other senior executives to do. I can, at a glance, view the performance of each of our managers and salespeople around the country. If I see something I don't like, I can fire off an electronic-mail memo. Conversely, if there is good news, I'm likely to contact the manager and congratulate him." That approach to management by walking around is quite a departure from the original intent and is limited by focusing on figures only. That was not the purpose of walking around. Looking at the figures does not tell the whole story. Providing a complex DSS to replace what should have been expected in this case as a need for immediate feedback from a well-qualified salesperson exemplifies an attempt to replace lack of competence with machines. The feedback that should have been immediate took 3 months; after installation of the DSS, it took 2 weeks. That improvement may indicate relative progress, but the information should have reached the appropriate individual before the end of the business day.

Cycle duration

In a generic sense, the cycle duration includes the time required to accomplish any task or activity independent of nature, size, or scope. That description applies to industry, educational institutions, government, and all organizational entities that attempt to meet some specific objectives. In its simplest form, cycle time can be described as the time it takes to meet some objective that begins at point A and ends at point B, as shown in Fig. 2.4. However, the designations of points A and B affect the final result. The achievement of this year's business objectives can be looked on as a single task or activity from the point of view of the CEO. The CEO's task is to lead the total organization in achieving the

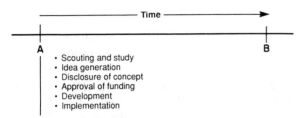

Figure 2.4 Cycle time starting point A.

business objectives. However, point A is not January 1 or the start of the fiscal year, and point B is not December 31 or the last day of the fiscal year. Point A is probably some years prior to this calendar or fiscal year.

The time from point A to B can be subdivided into many subtasks or activities. Whether the time to pass from point A to point B involves the concept-to-implementation phase of a new design for an automobile, a radically new computer system, or the time to specify the size and type of screw or electrical connector is not important. Each represents its own cycle time as well as some part of a larger and more comprehensive cycle time.

Establishing where points A and point B are provides totally different results. As an example, what activity determines points A and B for a new product development program? Thomas, as noted in Chap. 1, described them as when a customer expresses a need and when that need is fulfilled. But many business activities begin without knowledge of the customer's needs. The needs originate within the organization. That is especially true of long-term projects that begin with the exploration of new or emerging technologies. Figure 2.5 includes the more common terminology associated with decreasing cycle time. Generally, the focus is limited. The proponents of these approaches to simultaneous engineering take a short-term view. It usually relates to a specific product or process and not a series of products or processes—certainly not the total business.

1. Time to market
2. Enterprisewide development
3. Simultaneous engineering
4. Concurrent engineering
5. Concurrent process development
6. Research and development time
7. Product development time
8. Design time
9. Design and manufacturing time
10. Marketing and research time
11. Time to accomplishment

Figure 2.5 Terminology associated with approaches to decreasing cycle time.

Product concept to commercialization

Where is point *A*? In the majority of cases studied, it is the time of project approval. The time span from initial disclosure of a concept to formal approval is seldom considered. However, the period of time from describing a concept to the approval or funding of the project often exceeds the development time. It is essential that managers consider the investment in resources in the period prior to formal approval. There is no doubt that a great amount of research may be required to formulate a concept so adequately that it can be clearly delineated and presented for formal approval and funding. However, the time to decision plays a major role in cycle time optimization, and the time period from disclosure of a concept to formal approval must be managed. The activities in that period involve more than the originator. If the concept fits the objectives of the business, it should be evaluated and a decision about its importance to the business unit should be reached. The period is not the same as what is often considered to be scouting activity. Scouting precedes any clear exposition of an idea or concept.

The location of point *B* is equally important; Fig. 2.6 provides some possibilities. Point *B* could lie anywhere on the time line from disclosure of the idea to commercialization or implementation. Obviously, each of the activities mentioned in Fig. 2.6 operates within its own cycle and often in many sub-subcycles. But unless managers consider the comprehensive cycle time, that is, the time from disclosure of an idea to its commercialization or implementation, the exercise may deteriorate to game playing to see who wins the prize for reducing cycle time.

During a meeting with a group of managers, one participant provided an example of speed in introducing a new product. What normally would have taken two or more years was said to have been completed in 75 days. The new product development supposedly included the design of the product as well as the manufacturing process,

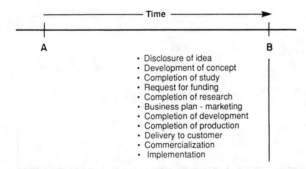

Figure 2.6 Possible endpoints of cycle time: point *B*.

the new packaging, and the field testing. Although such figures may appear to be impressive, they must be questioned. From concept to market readiness in 75 days may be possible for some minor modification, but not for a new product. If significant manufacturing processes must be designed and developed, they will not be developed, fabricated, installed, debugged, and a new product scaled up for production in 75 days. That type of game playing may be good public relations, but it does not add to bottom-line performance.

Any new product development cycle includes some variation of the following basic activities: scouting or study, idea generation, concept formulation, research, development, manufacturing, and commercialization. The same generic list of activities can be applied to transforming any idea, whether related to technologies, products, or markets, into reality. It is not limited to new product introductions. All of the activities apply, except perhaps *manufacturing* and *commercialization*. For those who object to the use of the words in relation to nonproduct related ideas, "processing" or "preparation" can be substituted for "manufacturing" and "implementation" can be substituted for "commercialization." So where should endpoint B be located in Fig. 2.6? The location depends on whether an organization functions as a holistic unit or a series of individual fiefdoms.

The following examples, which are not uncommon or unknown to most organizations, demonstrate the necessity for reflecting on the past and realistically appraising how an organization describes the limits of cycle time.

1. A proposal, with well-developed concepts, demonstration of the need for maintaining a competitive position, and financial and market justification, is presented for a major capital improvement of manufacturing facilities. There is a lapse of 5 years before a decision is made to proceed, and then engineering and its collaborators in manufacturing are expected to complete the design and construction in 18 months. Where was point A? In reality, it was 5 years before approval was given. Delays of that type do not apply only to capital projects; they extend to all the functional groups. The time to decision must be not only considered but emphasized in optimizing cycle time.

2. A global marketing concept is proposed for an extended and currently profitable product line with new products that has the potential for generating significant increases in revenue. Research, development, manufacturing, and marketing are committed to taking the risk, which appears to be moderate to low. But because of a lack of understanding of the customer's needs, the marketplace, and the benefits that will be provided by implementing the new technology, management delays the go-ahead decision for almost 3 years. Finally, the CEO or the decision-making group becomes aware of a continuing loss in

market share and begins pushing panic buttons. He or it now expects accelerated and aggressive commercialization and looks for all of the latest quick fixes to rectify a situation that could have been avoided.

Both of the preceding examples are continually repeated to greater or lesser degrees. The executives and the managers associated with each of the scenarios failed to recognize the negative consequences of delays in time to decision. They also ignored the impact on performance when people wait for decisions. Those who only stand and wait add cost.

Where do the various cycle times begin? Where, as an example, does manufacturing cycle time begin? My years of experience in managing research, development, and manufacturing operations clearly demonstrates that, at the very latest, manufacturing cycle time begins in the research laboratory. It may even precede the major expenditure of resources in research. It is too late to think about the most effective and efficient manufacturing processes when a product is in the final stages of research or development. The old processes may no longer be effective, or research or development may have spent additional time in an attempt to design products that fit the current manufacturing processes that may already be obsolete. Where does manufacturing cycle time end? It ends when the product is performing to the agreed-upon specifications.

The same question can be asked of marketing and sales. When does its cycle begin, and when does it end? It should begin long before the time at which a major laboratory or development effort is authorized. The marketing function should be a leading force in the description of the flow of new products needed in the marketplace. It should at least begin in the laboratory or in development stages, depending on the nature of the new product and the organizational structure. Marketing and sales input is not a matter of choice; it is absolutely essential in the early stages of any investment.

System Cycle Time Model

Optimization of cycle time requires consideration of many different business parameters. The model shown in Fig. 2.7 includes three distinct and interrelated components: the infrastructure, the business resources, and the types of activities. This model factors into the SCTM equation many contributing business operational parameters that affect cycle time. Too often, consideration of many of the elements is either disregarded as nonessential and irrelevant or totally ignored. The model expands the requirements for managing cycle time beyond increasing speed and relates the conditions required for taking full advantage of SCTM. In essence, managing cycle time lies at the core of

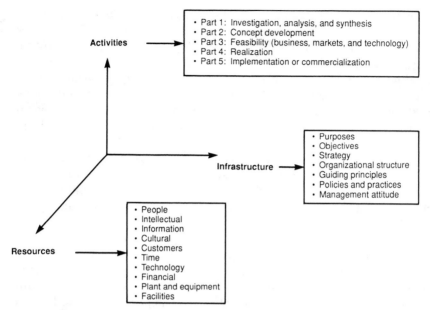

Figure 2.7 The system cycle time management (SCTM) model.

business management.

The SCTM model provides both a generic and a structured approach to review the critical issues that affect cycle time. Each of the three components of Fig. 2.7 includes specific issues that must be evaluated in relation to one another. That model will be referred to in subsequent chapters. The general classifications can be modified as required for any specific situation. Use of the model should not be considered as a prescription for managing cycle time. The model is presented to stimulate thinking about the essential elements of cycle time and their interrelations.

SCTM: Strategic or Operational Issue?

A question often raised is whether SCTM and the more limited interpretations of cycle time management are strategic or operational business issues. Without attempting to avoid making a judgment, the simple answer is that it depends. As a matter of fact, SCTM is both a strategic and an operational issue: It is strategic in the sense that it points the organization in a particular direction and with a well-defined business focus and that it demands high levels of proactive management competence. It is operational in the sense that the strategy is totally useless unless an implementation approach that involves the total organization is developed. A strategy without the associated implementation or operational plan that includes the allocation of the required resources and full knowledge of the limitations of the business infrastructure is just another irrelevant document generated by the bureaucracy. Strategy reflects some of the same mechanisms associated with new ideas and concepts: ideas and concepts without implementation provide no benefit. They need resources, a well-designed and developed infrastructure, support from those in places of decision, and, of course, the passionate determination of the initiators to bring them to reality.

As a practitioner, I have always emphasized operational issues while not minimizing or ignoring the need for a strategy. But strategy must take into account and be guided by unanticipated events that management may not be in a position to assess or predict with adequate precision [11]. Strategy has its place in presenting the overall framework of how an organization plans to meet its purposes and objectives. It provides a guidance system for the participants. It lays down some basic guidelines. Executives must make a clear distinction between strategic planning and strategy. The emphasis during the last decade on strategic planning was in most cases a waste of business resources. In spite of the rationalizations by CEOs who established the programs, the volumes of documentation provided neither strategies nor operational plans. So strategy must be placed in the proper perspective. By itself, strategy will not guarantee a flow of new ideas or products. A strategy that fosters innovation is useful only if it is backed up by an operational plan and the business infrastructure that allows innovative concepts to be implemented.

It is also necessary to be aware that attempting to implement a new strategy is probably most beneficial to the organization that first attempts it in its business sector. The followers will not receive the same benefits as the originator, and the strategy will be of lesser value to the originator once the competitors follow a similar strategy. As organizations look to other business sectors for input on new strategic directions, they must clearly understand the difficulties involved in managing a major cultural change. Managers must also understand that strategy is a moving target. What was acceptable yesterday may not be acceptable today. Strategy must be continually fine-tuned to the business environment.

Consider the strategies of 3M and Eastman Kodak in relation to innovation and entrepreneurship. 3M has a new product strategy that flows throughout the total organization [12]. It is backed up by management's support and the fact that it provides 15 percent free time to scientists and engineers to work on their own ideas. The company recognizes the need for innovation and the entrepreneurial approach in meeting its requirements that 25 percent of all sales revenue must come from products that were not available during the preceding 5 years. But the idea for continuous innovation and entrepreneurship has been ingrained in the 3M culture over many decades.

Contrast that situation with the attempts by Eastman Kodak to develop an entrepreneurial spirit. Kanter, in *When Giants Learn to Dance* [13], discusses Kodak's attempt, in 1984, to develop a more entrepreneurial culture or, as Kodak executives described the approach, to make Kodak "venture operative." Kanter states: "In order to nurture ideas that did not fall within traditional lines of business, Kodak set aside one percent of revenues for 'new ventures' which included a new venture process, acquisitions, and equity investments."

Kanter suggests that Kodak established a pioneering model that other companies flocked to examine. What happened to this "venture operative" attempt? Hirsch, in *The Wall Street Journal* article headlined "At Giant Kodak, 'Intrapreneurs' Lose Foothold" [14], describes Kodak's attempts to institute an entrepreneurial spirit within its monolithic structure. Of the 14 ventures that Kodak established, 6 have been shut down, 3 have been sold, 4 have been merged into the company, and only 1 operates independently. The difficulties that created this situation are not uncommon. Kodak, over the many decades, developed a very wide comfort zone for its employees. It had no serious competition. It was king in the photographic business and without any global competitors—it controlled the marketplace and controlled the introduction of new products. Under such conditions, an organization loses sight of its costs and, perhaps even more detrimental over the long term, loses its expectations of performance by all of its participants. Words like "stretch" and "push" were probably eliminated from the company vocabulary. As one of the entrepreneurial leaders at Kodak mentioned, he supposedly was in control, but he was forced to use Kodak engineers, researchers, and other employees in joint-development efforts. He also noted that "The Kodak people would take their eight-hour workday and go home." Entrepreneurial or innovative organizations are neither built nor sustained when the participants begin to watch the clock.

Although one of Kodak's strategies for growth may have been directed toward developing the entrepreneurial spirit, it resulted in something far less than anticipated. This is not to say the strategy was wrong, but somewhere along the line Kodak executives did not understand the requirements for developing a culture that would foster entrepreneurship. Attitudes required changing. 3M's entrepreneurial attitude is ingrained in the organization. 3M is a diverse multiproduct business that not only provides opportunities to its employees but expects its managers to support those activities and function as sponsors. Whether a 3M type of culture can be transferred to an organization like

Kodak in 5 short years is debatable. Eastman Kodak is a well-respected organization, but it never functioned as an entrepreneurial enterprise. The situation of Kodak is not uncommon. Citing the Eastman Kodak case demonstrates the difficulties of creating change in a highly bureaucratic enterprise. The problems associated in this situation did not originate because of a lack of technical or marketing talent. The problem lies in the fact that the executives lack the understanding of what is required to change a culture. A summary statement to the Kodak management and their entrepreneurs might be that the infrastructure as described in the SCTM model must be in place. That infrastructure requirement relates not only to managing cycle time but also to managing the business.

Shortening or Optimizing

Most of the publications related to simultaneous engineering as approaches to managing cycle time (see references in Chap. 1) emphasize the reduction or compression of cycle time; they imply that faster is in some way better. Speed seems to dominate the thinking. But as with so many other activities, maximum speed may not be the answer. The question that really needs to be asked is this: What is the appropriate speed at which to accomplish the particular objective? "We have to shrink the time schedule" is often heard during product development. But that time can be reduced only if all of the resources are available and the infrastructure is in place. Merely reducing the schedule on paper will not make it happen.

There is little value in reducing the cycle time if the finale develops into a hurry-up-and-wait situation. That approach only demotivates and adds cost. Speeding up a project beyond certain limits adds cost. It usually involves assigning more people than would normally be required, more coordinators, more expediters, and so on. In the attempts to speed up operations, various types of technology and market audits that cause future problems that must be resolved are omitted. Compromises in design lead to production problems that add to future cost. Likewise, design compromises lead to major product modifications with future limitations for enhancements. There is nothing wrong with the concept of speed, but like the driver of an automobile, the manager cannot exceed the speed limits, go through red lights, cross the median, and so on, and hope to accomplish the objective in an optimal manner. Speed can be an advantage if, in the process of acceleration, all the essential steps are followed and the available resources are utilized effectively and efficiently. The process must be followed unless there are some verifiable reasons to modify it. As an example, a relatively normal and essential activity such as formal documentation lags the work effort significantly and generates countless hours of nonessential argument, disruption, changes in tooling, manufacturing, and so on, at greatly added cost. Such activities add time and cost—total cycle time as well as hours for accomplishment.

Optimization presents a more workable and logical approach. Viewing SCTM through optimization glasses allows the participants to organize their work effort flexibly and speed up certain subproject activities and slow down others as required. As an example, assume that a project is made up of 100 very specific activities. An activity could involve only one person or a group of persons. It could also include the interaction of two or more activities depending on how the project is subdivided. To optimize cycle time, it may not be necessary to speed up all the activities. Some activities may actually be stretched out over a longer time period, but that does not mean an increase in the total number of hours. Only the activities that are essential to meet the new requirements will be speeded up. Optimization in that sense can eliminate or reduce the added costs that are generated when *speed* becomes the dominant variable. If an organization attempts to improve manufacturing productivity by 20 percent, it does not succeed by merely speeding up all of the machines or operations by 20 percent. That would lead to chaos, and many machines would probably break down in a short time. Each element of the speedup must be evaluated individually to determine the potential as well as the need for the increased speed. Some study would be required to develop a workable plan. Speed may dominate optimizing cycle time, but the objective must take into account the total hours. Doing the same job in the same way, but faster, may be the wrong approach. The job content must be reevaluated, and a new job design must be developed to gain a benefit.

Any business activity involves the allocation of resources. There are always tradeoffs that can be made to optimize the total use of resources. Too often, managers consider people, finances, plant and equipment, and facilities as the only resources available. Furthermore, they generally limit the available resources to those that are accessible within the organization. They forget that resources are available nationally and globally. Ideas, concepts, discoveries, and talent are not limited by the physical boundaries of the business unit. Optimization places no limits on the sources of the required resources. It provides for substitution of one type of resource for another based upon the projected benefits.

Optimization changes the mindset of the organization. Like the manufacturing process that is optimized rather than speeded up, it allows for a choice of which parameters will operate at what speeds and under which conditions so that the system as a whole, rather than one or more individual parts, will be optimized. It functions most effectively when each component or element operates at the right speed, is within its established limits, and supports every other component or element.

SCTM reflects all of the attributes of thinking in terms of systems rather than discrete parts. Optimization of the system takes precedence over maximizing the output from any one function. Although the actions of the discrete parts must be directed toward performing certain specific activities, those actions must take into account their relations to the total system. That approach in no way reduces the need for managing the parts, but the parts are managed to optimize the system. An exceptional R & D department that became a reality as a result of hiring creative and innovative scientists and engineers is of little value if it is not supported by similar levels of competence in manufacturing and marketing. A similar statement applies to the subdivided elements of research. When a competent group of polymer chemists in R & D is not augmented by an equally competent group of process engineers, the potential of the total system is limited. An efficient manufacturing organization provides little benefit if the flow of new products from research and marketing is lagging the competition. Within manufacturing, an ineffective order entry system limits the potential efficiency of the manufacturing function. SCTM requires the active support and the integration of the activities of the functional groups in order to optimize the output of the system. Those comments may appear to be elementary, but the management press continues to report negative business results because organizations have failed to emphasize the need for optimizing the system.

Summary

Managing cycle time cannot be treated as a single issue or a quick fix. It is important that managers understand that an organization manages its cycle time only when managers broaden their perspective beyond their immediate functional responsibilities and focus attention on the business objectives. That approach requires significant changes in attitude both by managers and senior executives. Chapter 2 described the many different aspects of managing cycle time. There are no prescriptions to follow, just attention to the details and practicing the fundamentals of management. Managing cycle time demands the best in the attitude with which managers approach the practice of managing. It is time-consuming, and it requires a new perspective not only by executives, managers, and specialists but by the total organization.

Figure 2.8 summarizes the three components of cycle time. Type, class, and the three elements of cycle time interact in an effort to gain the benefits of managing cycle time. As an example, a system cycle is bound by all of the decision classes and the three elements. None can

Components of cycle time		
Type	**Class**	**Elements**
• System • Product service process • Project	• Predecision • Decision • Decision to activity • Activity • Activity to implementation	• Time • Timing • Cycle duration

Figure 2.8 The components of cycle time.

be omitted. By systematically going through the activities listed in Fig. 2.8, managers can go beyond the cursory discussions related to cycle time and determine where time can be managed more effectively; the emphasis must be placed on operating at the right speed. Time compression for individual activities provides many benefits, but only if that time compression benefits the total system.

Gaining the benefits from SCTM requires proactive managers. Each cycle consists of many related and interconnecting cycles that must be analyzed and synthesized into a workable solution. By describing the many cycles involved in reaching an objective, managers begin to understand why cycle times extend beyond some acceptable limit. The process allows managers to manage the system cycle rather than be managed by it. Identifying the starting point and the endpoint of the cycle forces managers to think beyond the prosaic and focus attention on innovative approaches to the management process.

The system cycle time model takes into consideration the three components and their interrelations in managing the components of cycle time. In that sense, managing cycle time embraces all the fundamentals of management.

References

1. G. H. Gaynor, *Achieving the Competitive Advantage through Integrated Technology Management,* McGraw-Hill, New York, 1991, pp. 13–15.
2. G. H. Gaynor, *Achieving the Competitive Advantage through Integrated Technology Management,* McGraw-Hill, New York, 1991, pp. 19–41.
3. L. R. Bittel, *Right on Time!,* McGraw-Hill, New York, 1991.
4. Jean Claude G. Usunier, "Business Time Perceptions and National Cultures: A Comparative Survey," *Journal of International Business,* vol. 31, no. 3, 1991, pp. 197–212.
5. E. T. Hall, *The Dance of Life: The Other Dimension of Time,* Anchor Press/Doubleday, Garden City, N.Y., 1983, pp. 13–54.
6. D. McGregor, *The Human Side of Enterprise,* McGraw-Hill, New York, 1960.
7. W. Oncken Jr., *Managing Management Time,* Prentice-Hall, Englewood Cliffs, N.J., 1984, pp. 2–18.
8. R. H. Beeby, "How to Crunch a Bunch of Figures," *The Wall Street Journal,* June 11, 1990.

9. R. J. Chambers, "Don't Let Computers Do All Your Thinking," *The Wall Street Journal,* letters to the editor, July 11, 1990.
10. T. J. Peters and R. H. Waterman, *In Search of Excellence,* Warner Books/Harper & Row, New York, 1982, pp. 121–125.
11. G. H. Gaynor, *Achieving the Competitive Advantage through Integrated Technology Management,* McGraw-Hill, New York, 1991, pp. 69–86.
12. G. Pinchot III, *Intrapreneuring,* Harper & Row, New York, 1985, pp. 32–64.
13. Rosabeth Moss Kanter, *When Giants Learn to Dance,* Simon and Schuster, New York, 1989, pp.33–43.
14. J. S. Hirsch, "At Giant Kodak, 'Intrapreneurs' Lose Foothold," *The Wall Street Journal,* Aug. 17, 1990.

3

Identifying the Origins
of Lost Time

Managing cycle time requires managing time, but not in the traditional sense related only to efficiency. Cycle time must be managed from a business systems perspective. Cycle time management is a matter not of reducing hours for completion of an activity, but of using an optimal number of hours—the hurry-up-and-wait approach provides no benefits. It is not a matter of reducing talk around the water cooler or the coffee machine—there can be some profitable discussions at those nonworking locations that seem to obsess managers. It is not a matter of restricting freedom to think the unthinkable or propose seemingly impractical solutions—identifying the origins of lost time extends beyond such simplistic concepts. It is not a recreation of Charlie Chaplin's *Modern Times* for managers and professionals. Chapter 3 guides the reader in identifying the direct and the indirect, the internal and the external sources of lost time:

- Starting at the top
- Originating in the system
- Created by management
- In the functional groups
- Frittered away by people
- The lost-time cost report

Lost Time—Starting at the Top

To manage time and its impact on cycle time in technology management, the origins of lost time must be identified and acknowledged at all levels of management beginning with the CEO. Convincing managers to acknowledge the fact that certain activities of their own or those that they impose on others do not add value to the organization presents a major management challenge. Lost time is a waste of resources. In today's economy that level of waste determines business performance. The effect of lost time originating at the upper levels of management increases geometrically as it passes through every level of management. As an example, a delay in a decision at the executive level can result in thousands of hours of lost time or wasted effort. If an organization decides to manage cycle time, the origins of lost time must be identified, acknowledged, and eliminated if possible or at least substantially reduced.

Under the heading "Seeing Red over IBM's Blues," *USA Today* [1] reported that John Akers, the CEO of IBM, remarked that "employees are too damn comfortable at a time when the business is in crisis." Akers supposedly let loose at IBM managers and employees about the company's unsatisfactory performance. Some pertinent questions must be asked. Why were managers and professionals allowed to become so comfortable when sales and profits were not meeting expectations? What were the results of all of those staff activities that track competitors' technologies and markets, measure department productivity, and get involved in unrelated business activities? Why did senior management ignore the handwriting on the wall? In a follow-up article in *Business Week,* "IBM: As Markets and Technology Change, Can Big Blue Remake its Culture?" [2], Akers says, "The fact that we're losing share makes me goddamn mad....The tension level is not high enough....The business is in crisis." Akers must have played a role in developing that culture. *The Wall Street Journal* (November 27, 1991) [3], announced "IBM Plans $3 Billion Charge and About 20,000 Job Cuts" in 1992, bringing the total employment to about 325,000 from a peak of 407,000 in 1986. On December 2, 1991, *The Wall Street Journal* [4] in an article "How an IBM Attempt to Regain PC Lead Has Slid into Trouble" relates IBM's management problems:

> It shows IBM's bureaucracy at its most arthritic and how Big Blue throws billions of dollars at problems that can't be solved with money. It shows how IBM clings to the idea that it must control every part of the market, even though the company must relinquish that heritage if it is to be as flexible as Mr. Akers' latest plan requires.

This is the company that Peters and Waterman [5] included in their list of excellent corporations. It met all of the requirements: a bias for action, close to the customer, autonomy and entrepreneurship, productivity through people, hands-on and value-driven, stick to their knitting, simple form and lean staff, and simultaneous loose-tight relationships. It also met the financial criteria. IBM is cited in over 30 specific references. Perhaps IBM executives and managers took the comments of Peters and Waterman too seriously. They thought they had achieved excellence and failed to recognize that excellence is a moving target. They accepted the Peters and Waterman comments as gospel without reflecting on the reality. Companies that develop into quasi-monopolistic entities in a single major industry tend to be blinded by success. Lack of effective competition breeds neglect of the fundamentals of managing. Effectiveness and efficiency are not considered to be essentials of managing. Such organizations as General Motors and Eastman Kodak have faced similar problems. When a single business dominates an industry, it neglects to pay attention to the factors that brought it into being and allowed it to prosper.

Changing any culture involves some pain. It also consumes many hours of unproductive time. Changing the IBM culture from one of self-satisfaction and complacency to one of energy and drive essential for competing in today's global markets first requires executives to change themselves and recognize their fiduciary responsibility to the stakeholders. How IBM and Akers will manage the future is uncertain. Bureaucracies are difficult to change. IBM is not the only organization that faces such problems. The virus has infected most U.S. businesses and the whole nation, and yet the executive suites have not responded to creating the necessary change in attitude. The problem really needs some creative thinking by the CEOs. They must get involved and consider the importance of effective and efficient use of time on business performance. Downsizing is not the magic answer. What counts is what the CEO does with what remains after the downsizing. If the incompetent remain and the same attitude prevails, there is little chance of developing a proactive organization.

Digital Equipment Corp. is confronting difficulties similar to those at IBM. A *Fortune* article [6] reports on how CEO Ken Olsen, the company founder, is attempting to move an organization of almost 125,000 people toward more productive performance. The company not only lost out on the powerful workstation technology; in the process, it lost touch with its customers. A look at the sales per employee for 1990 shows Digital with the lowest productivity performance when compared to Hewlett-Packard, IBM, Sun Microsystems, and Apple. While sales per employee must be viewed carefully, there is sufficient disparity to indicate that either a bureaucracy has taken over or the organization has eliminated the word "productivity" from its vocabulary. Olsen plans to bring Digital back to its onetime preeminent position with a *combination of basic engineering and brute force.* The no-layoff policy of the past, similar to IBM's, developed a comfort zone that cannot survive in a competitive environment.

As with IBM, Digital must change its culture to one that focuses attention on employees who contribute to the development of the business. It must focus attention on the use of time for adding value. That change must begin with executives and managers who are responsible for the current situation, which they created. Management has a responsibility for more than just keeping the doors open. It must anticipate business needs and recognize the impact of emerging technologies, changing markets, and the other competitive forces that dominate its particular business. In spite of Digital's people orientation, there is a discordant note in the comments of Olsen, who states: "People are always reluctant to change. Teasing them, tricking them, manipulating them—that's all part of the job of management." Serious questions must be asked as to whether such an operating philosophy will provide the impetus for the major changes in employee attitude that are required. Employees might seriously question the applicability as well as the integrity of Digital's human resource policies.

The examples of IBM and Digital, two well-respected companies, demonstrate what occurs when management loses touch with the realities of managing the business. Certainly the companies can be commended for their emphasis in the past on respect for the individual, but that respect must be supplemented by raising the performance expectations. Respect for the individual cannot be translated into acceptance of mediocre performance or just plain lack of performance and a lack of business discipline. Respect for the individual also includes providing guidance so that the individual is not allowed to vegetate. Perhaps the human resource and the training professionals taught from the wrong textbook. Respect for the individual is essential, but by itself is insufficient for long-term success in a competitive marketplace. It requires a response from the individual and acceptance of responsibility for performance at the highest professional level.

What have these two examples to do with managing cycle time? They demonstrate that lost hours affect performance negatively now and for some time in the future. There is no way to determine the lost hours. But if each employee at IBM lost only one hour as a result of Akers' comments to management, that alone would generate over 350,000 hours of lost time. The difficulty is that time lost or wasted by executive comments or actions is not factored into a lost time report. CEOs and top-level executives continue to think that they are the only people who discuss such matters. They prefer to talk about the lost time that occurs at the coffee machine. Time *does* play a major role in business performance if the value of time is expressed in meaningful terms. Determining the *cost of lost time* should awaken executives to the potential financial benefit and the positive impact on people performance. They must recognize how their communications and actions generate inordinate amounts of lost time. The figures will exceed the gross profits of most profit-making organizations.

Lost Time Originating in the System

Figure 3.1 provides a model for identifying the four origins of lost time: the system, management, the functional operations, and people, including executives and managers. Some of the lost time is attributable directly to those four origins and some indirectly to the actions of others. Also, some lost time can be charged to internal as well as external operations. Each of the four origins is described by the direct effect on performance and from the internal perspective of the organization. The affecting factors, indirect influence and external influence, on the four origins are considered separately because they cut across all four.

The system—meaning the manner in which the business is organized, the policies and procedures that guide operations, and the communication processes—either enhances or detracts from business performance in proportion to the lost time it generates. Executives seldom consider the lost time generated by their actions or their comments. How many lost time hours will be allocated to the comments of Akers of IBM and Olsen of Digital? For every employee at least an immediate hour of discussion will be followed by many additional lost hours until the system settles down at some new steady-state condition, which will be at some depressed level of performance. The CEOs are trying to change the systems without preparing the participants for the change; the change is from the top rather than the top, the bottom, and the middle.

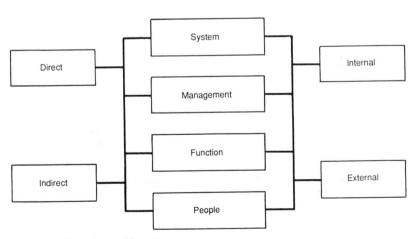

Figure 3.1   The origins of lost time.

Some examples of lost time generated by the system will help an organization begin the task of recognizing areas for improvement. System lost time arises from:

- Guiding principles
- Policies
- Procedures

Dividing the system lost time by using those three categories provides a means of identifying and prioritizing potential areas for improvement.

Guiding principles

Every organization is guided by principles that may or may not positively affect performance. As organizations grow and prosper and business conditions change, the principles that once fostered success become stumbling blocks to progress.

Bureaucracy. There is no lack of criticism by business of the bureaucracy at all levels of government, but business is rapidly approaching the same levels. Bureaucracy is necessary, but it must be efficient and meet the needs of the organization. Employees want their paychecks on time; vendors want to be reimbursed on time; and customers want orders delivered on time. Although creativity and innovation also are essential to bureaucracy, the length of time between changes in the system are long compared to product life cycles.

Responsibility and accountability. Responsibility and accountability at all levels of the organization receive little if any attention from top management. In general, people do not take responsibility for performance. Personal responsibility appears to have exited businesses in the late 1960s. At one time, if people, either individually or in groups, made a commitment, that commitment was fulfilled, even if it meant working overtime at personal cost and without additional compensation. The attitude was that even though overtime was not directly compensated, the person or group would be rewarded in some way for the extra effort. That attitude is seldom found today. It is only necessary to look at the industrial campus parking lots after 5:00 PM or on weekends to realize that most work ceases at 5:00 PM on Friday and begins at 8:00 AM on Monday. What does that attitude cost a business and what does it cost the nation in lost productivity? How does it affect the profitability of a product that should have been introduced 3 months ago? All this is not to suggest that everyone should be expected to put in additional hours without compensation, but when those hours are needed to fulfill a commitment, they should be expected. Businesses involve adults, and adults and particularly managers and professionals should meet their commitments as professionals.

Welfare mentality. Executives decry the development of the welfare state—meaning the government. Yet in reality, many businesses have, to various degrees, generated their own welfare states, beginning in the executive suites. Many executives complain about government entitlement programs, but they have developed their own. A welfare mentality has been developed: A business not only hires an individual to make a contribution to the business but takes on all the personal problems of the individual and his or her family as well. Businesses provide free continuing education, counseling in financial matters, guidance for chemical dependency, facilities for day care, and so on. All those services emphasize concern for well-being of the employees, but productivity of the white-collar work force continues to decline. Business executives must take the responsibility for creating the welfare attitude. The reason why the situation arose is quite simple: lack of management involvement. Business executives, not being literate in the requirements for managing human resources, delegated the decisions to the human resource professionals, who accepted the theories of the behaviorists and other academics without any evidence that the theories really work.

Education. Industry spends billions of dollars in education and training annually. How much of that education translates into improved business performance? The simple answer is, not very much. IBM maintains one of the best industrial educational centers for employees and managers in the world. If industrial education is providing a benefit, why does IBM find itself in its current position? Perhaps too much education centers around the myths and heroes of the past. There is no doubt that education, in an economy in which 50 percent of the learning of a professional specialist may have little value after 5 to 10 years requires some type of continuous effort. There is also no doubt that investment in education must provide a benefit to the organization. There is no way to determine specifically the net worth of additional management education, but learning new concepts, approaches, or techniques and methodologies can be measured by the number that are introduced effectively and the length of time required to introduce them. As an example, education in the fundamentals of managing quality have taken on important significance in the last 10 years. The results of that education can be measured. The question that must be asked for future programs is this: How much of that education was really required? There is no doubt that organizations spent more than was necessary.

Culture. Organizational culture determines how time is used and whether it is considered to be a limited or unlimited resource. It is one of the determining factors in cycle time. If culture perpetuates the myths and overemphasizes the past, it affects cycle time negatively. It can be a positive force if it provides the behavioral changes required to optimize cycle time. Culture involves being people-oriented, but people-oriented to the extent that competence is adequately challenged. It requires developing an environment that fosters innovation and balances conflict and diversity with the organizational purposes, objectives, and strategies.

Policies

A clear distinction between policies and procedures must be made. Policies can best be described as guidelines that establish some limits of operation for all the participants in the organization. They apply to everyone and not just to managers, as is often incorrectly assumed. They involve people relations, human development, career emergence, introduction of products, managing quality, satisfying customers, changing behavior, establishing supportive community relations, providing educational opportunities, meeting financial requirements, and all of the individual policies of the functional groups.

Human resources. Suppose a company policy that provides exclusively for promotion from within; no managers are hired into the company. Although that policy provides opportunities for people within the organization, it also deprives the organization of outside expertise and the benefit of developing change. It resists change and often emphasizes half-truths. It can breed managerial and operational waste. The following statements are common in many organizations:

- That's not the way we do it around here.
- Top management doesn't care for that kind of approach.
- That kind of thinking is dangerous around here.
- The guys at the top are on the board of company so and so.
- Headquarters says—but the individual is not identified.

Such statements create delays by introducing uncertainty and the follow-up need for verification. First, chatter or noise of that type wastes time. Second, it creates an attitude of mistrust and cynicism that diminishes the output of the total organization.

Technology. The policies for dealing with the many different aspects of technology can enhance or detract from business operations. A policy that emphasizes development of technology solely within the organization may consume resources that could be allocated more effectively if the technology were purchased or some form of joint venture were established. The not invented here (NIH) syndrome permeates all organizations at a significant cost of time and money. Although innovation determines the future of most organizations, it too must be balanced by the benefits provided. Not all innovations are worthwhile. An innovation policy must take into account the resources and the infrastructure elements discussed in Chap. 2.

Quality. There are relatively few organizations without a policy regarding quality, and generally the focus is on production. To be of any value, however, a quality policy must focus on every activity and every person in the organization. Whether that policy affects the quality of the product or service being marketed positively depends on top-management's approach—real improvement or public relations hype. Quality, in the final analysis, is the judgment of the customer. To a great extent, it depends on how well the customer has been educated. Although top management prefers to allow the customers to think they are always right, the reality is somewhat different. Customers must be educated. They have no pact with the seller. They may or may not know whether they are receiving a quality product. Inadequate quality adds cost and lengthens cycle time.

Intellectual property. Intellectual property is an asset and is one of the business resources that must be optimized. Too often, individuals entering an organization are not privy to the vast storehouse of knowledge that may have been accumulated in the past. Taking advantage of that intellectual property and exploiting it to the fullest can reduce cycle time. Often, too much time is spent looking for the major breakthrough in some off-site location instead of taking advantage of past experiences and information already in existence. Ignoring the organization's intellectual property leads to lost time and extended cycle time.

Procedures

Procedures describe the process by which certain activities are performed. The degree of specificity of the description can help or hinder the process. There is no doubt that procedures are essential, but they cannot be reduced to recipes or prescriptions. They must be flexible and make allowance for personal judgment when conditions necessitate deviating from the ideal. Judgment is essential; otherwise, an organization atrophies. Computers are not going to eliminate our thinking processes. For purposes of consistency, examples of procedures will relate to the previously mentioned policies related to people, technology, quality, and intellectual property.

Human resources. There is no lack of procedures related to managing human resources. The continuum includes all the possibilities from the procedure for determining the needs to the procedure for terminating an employee. Procedures for bringing new people into the organization represent the greatest amount of time, that is, the time from advertisement or first contact to actual participation in some meaningful activity. Managers usually underestimate the time it takes to find the specific person with the necessary education, experience, and other attributes. Furthermore, too many managers do not even know how to communicate specific needs adequately. The net effect of any ambivalence is extension of cycle time. As an example, if a project schedule is based on the availability of specific expertise and that expertise is not available in a timely manner, time will be wasted and cycle time will be extended.

Technology. Procedures related to technology extend to many different operations. They involve the search, the negotiation process, and the acquisition of new technologies. They can relate to the justification for investing in existing or new technologies related to research, development, manufacturing, or any of the other functions of the organization. They can relate to procedures for exiting certain technologies. Valuable time can be lost in strictly following each of these specific procedures. As an example, the justification procedures that normally require many levels of approval can take many months for completion. There are further complications if the approval originates in a foreign subsidiary. In such cases, the home office usually just accepts the fact that nothing can be done to reduce the time of approval except to have the subsidiary begin the process earlier. Reducing project approval time begins with a complete reevaluation of all the elements in the approval chain. It includes the amounts that can be authorized at various levels of management and acceptance by managers at those levels of responsibility and accountability for their decisions.

Quality. Procedures drive most quality issues: to allow deviations from the specifications or to hold to specifications rigidly, to purchase from one vendor versus another, to continue production or to stop production, to ship or not to ship, and to service or ignore the concerns of the customer or the user. Those are major quality issues, and debating them every time they become apparent only loses time. An organization must provide some procedures for handling deviations from specifications or performance anywhere in the research to customer process. Companies that continue to manufacture out-of-specification components, assemblies, or products with the expectation of rework only add to their costs and their total cycle time. If procedures are established and then violated by top management or are in any way compromised by technicalities, the time that was allocated to establishing the quality-related procedures is wasted. The current mania sweeping the executive offices in competing for the Baldridge Award [7] forces one to question whether executives are playing games with a most critical element in the global competitive equation. Process and procedure are essential, but they do not guarantee quality products or services. *Business Week,* in "The Ecstasy and the Agony" [8], raises some issues regarding preoccupation with the Baldridge Award. Wallace Co., a 1990 recipient, is struggling against a tide of red ink. In the process of achieving the award, overhead increased $2 million per year and customers balked at price increases. In 1990, the company lost $691,000 on revenues of $88 million.

Y. K. Shetty, in "Product Quality and Competitive Strategy" [9], raises some questions whether the past efforts have made American products significantly superior. In a study of 171 companies, quality was ranked sixth among a class of productivity improvement programs. In the same article, J. Harrington, IBM quality manager, says that about 25 percent of manufacturing and administrative *time* is diverted to repairing defects and correcting errors. He estimates that eliminating these losses could increase output by more than 25 percent. Reducing the cost of quality leads to gains in productivity and profitability. Time is an important factor. Time lost in correcting errors is another example of where time is wasted.

Intellectual property. Procedures for managing intellectual property are relatively new. Not too many organizations are as yet totally aware of the value of their intellectual property. They have not devised the means for exploiting it, because their environment does not support the innovation process and is bound up in narrow parochial beliefs. Exploiting the intellectual property reduces lost time, improves cycle time, and builds new businesses. Post-it notes [10], developed by 3M, are an excellent example of what can occur when time is taken to review technologies and laboratory results that were relegated to the archives. Several major procedures are essential for managing intellectual property in order to capitalize on the valuable resource:

- Understanding its importance and value
- Maintaining an inventory
- Providing easy access
- Auditing on a periodic basis
- Forcing the use if necessary

Time Lost by Management

As a group, managers contribute to time lost throughout the organization, often without recognizing the implications of their requests. Competence and management are not synonymous. Managers cannot blame the rank and file for nonperformance. As an example, somewhere in the process of climbing the corporate ladder at IBM, Akers [11] contributed through his decisions to the current dilemma facing the company. He helped to create the environment. Whether he acceded to the wishes of others or was an active proponent of developing the comfortable lifestyle is unimportant. He cannot expect to change the embedded lack of energy by a get-tough policy of words with the managers.

Managers must be reeducated to understand their roles in business performance. The psychological gurus have spread the word that managers accomplish goals and objectives through others. They forgot to include one additional term in the management equation: the contributions expected from the manager. Top management bought the approach and rewarded it. Too many managers have been recipients of the corporate welfare check for the last two decades. Their direct contributions to the business seldom equate to the levels of their compensation. In the process they either knowingly or naively extend the cycle time. The list of activities in which managers consume time, in which no value is added as a result of the activity, could be endless. Some examples, without prioritizing, may help bring managers to a realization of how they extend cycle time.

Planning

The link between strategic planning and performance has not been established. For several decades, academic researchers and some managers have been attempting to find a positive correlation between strategic planning and performance. Both sides of the issue have been debated. All types of strategic planning have been analyzed. Even the intangible benefits have been considered. V. Ramanujam and N. Venkatraman, in "Planning and Performance: A New Look at an Old Question" [12], attempt to relate the differences in planning practices between "high performers" and "low performers." They consider the critical role of planning as the ability of the system to support strategy development and implementation. The roles that determine the system's effectiveness include the system's ability to anticipate surprises and crises, its contribution to management control, and its ability to enhance creativity and innovation. They conclude that good planning requires resource commitment, needs line-staff cooperation, requires retrospection, is built on functional integration, extends beyond techniques, and does not involve a trade-off between creativity and control.

The conclusions of Ramanujam and Venkatraman involve time. Good planning must take into account the effectiveness and the efficiency with which it is accomplished. Too often, excessive strategic planning leads to no strategy. The cost in time of the volumes of data that contain neither a strategy nor sufficient operational details must be determined. Planning is essential, but it must be accomplished by those who will be expected to implement the plan.

Competitiveness

Competitiveness has received and will continue to receive much attention from both the academic and the business community. In recent years, managing information, innovation, quality, technology, and many other approaches have been presented as the means for becoming globally competitive. They have been presented independently as panaceas for neutralizing competitive pressures. There is no doubt that each of the programs affects the ability of an organization to compete, but by itself any one program is an exercise in futility. As an example, managing information is important, vitally important, but it must be supplemented by comparable effort throughout the organization. The most sophisticated and technically advanced management information system provides little benefit if the system lacks the capacity for innovation in all aspects of the business. Implementing that magic information system may have been a total waste of time.

Information transfer

Information transfer ranks high on the list of time wasters. Expecting too much information or insisting on "in depth" studies, when something less would be sufficient, consumes valuable time. Gathering "just in case" information that may only satisfy the whim of some executive also consumes time. The guarded transfer of information between the functional groups not only wastes time but delays progress. Classifying information that is public knowledge as "confidential" or labeling strategic plans as "for executive management only" exacerbates the information transfer problem. Some questions need asking: Do I need the information to do my job? What will I do with it and how will I use it ? How can it affect my overall performance?

Using coordinators

Use of coordinators not only wastes resources but tells the story of how a business functions. Organizations appoint managers to take responsibility and be accountable for certain operations. Then they appoint coordinators, often at higher levels, to coordinate various functions. The latest example that comes to mind is the coordinator of quality programs. The individual should either be responsible for implementing the quality program or not be involved. Occasionally there may be a temporary need for a coordinator in some operation, but if that need persists, it is time for management to determine why one is needed.

Changing priorities

To change priorities consumes both direct and indirect hours of time throughout the organization, especially if these changes are frequent. There is no doubt that priorities will change, but the degree to which change undermines business productivity must be controlled. Changing priorities at all levels and in all the functional operating groups has the same negative effect. A further negative effect is experienced when those changed priorities are not communicated in a timely manner and to all of the people involved.

Feedback

Inadequate feedback to the organization about all matters related to the business leads to time lost by speculation about the unknown. Whether the topic is organizational change, business performance, some unexpected event, or general economic conditions, time is spent on theorizing and speculating about outcomes. This phenomenon usually begins at the top of the organization and moves throughout the system.

Inadequate direction

Lack of a clearly enunciated vision and the attendant purposes, objectives, and strategies causes countless hours of time to be lost. Like other impediments to the effective use of time, vision, purposes, objectives, and strategies must be translated so they will have meaning at the many different levels in the organization. They must be interpreted so they will have some substance. If the rank and file cannot, so to speak, touch and feel them, they are most likely useless pronouncements. The vision for the future must be consistent with executive actions. Purposes must be specific, even though broad, to focus organizational effort. Objectives cannot be stated in esoteric financial terms only: They must indicate to the employees what the company expects from them and how they will participate. The strategies for achieving the vision, purposes, and objectives must be interpreted within the individual subcultures that exist in every organization. They must be strengthened by appropriate management attitudes and adequate resources for meeting future business requirements. Without clear and specific enunciation of business direction, time will continue to be wasted.

Illiteracy

Illiteracy of all types and descriptions reduces the time available for productive activity. A great deal of illiteracy permeates business in all disciplines. It is not limited to the "computer illiteracy" that makes headlines. It includes business, technology, marketing, political, societal, international, and many other types of illiteracy. The lack of business literacy, that is, what the business we are in is all about and how our business fits into our industry, permeates all organizations at all levels. How many individuals beyond those working in finance have an understanding of the related financial requirements? How many understand the fundamental issues in cost accounting? Business illiteracy prevails among many technical managers in research, development, and manufacturing who take a narrow view of their responsibilities and disregard the importance of just how they fit into the system. Technological illiteracy abounds outside the technically oriented functions. Marketing illiteracy causes a similar loss of time. Executives and managers cannot be technological and marketing professionals, but they must have a minimal level of understanding.

Other system-imposed time wasters

The list of time wasters generated by management is long; Fig. 3.2 is a partial list to jog the memories of managers who plan to deal with the importance of time in SCTM. Every organization must develop its own list of managerial malpractices that are consuming valuable time, not providing any added value, and negatively affecting timing and cycle time.

1. Ignoring the importance of up-front thinking
2. Lacking leadership in creating and managing constructive change
3. Mismanaging the risks and uncertainties
4. Procrastinating in decision making
5. Tolerating noninvolved and nonparticipatory managers and professionals
6. Perpetuating the imbalance between credentials and prior performance
7. Ignoring the need for organizational and management discipline
8. Managing by fads, slogans, quick fixes, and single issues
9. Taking a myopic or narrow view rather than looking at the system
10. Playing games instead of managing businesses
11. Mismanaging the long-term technology benefits by solely emphasizing the short-term results

Figure 3.2 Partial list of time wasters to jog the memories of managers attempting to control system cycle time.

Time Lost by Functional Groups

Functional groups generate their own lost time primarily because of the manner in which they work, and that is in addition to the lost-time activities imposed by the system and management. Each function operates according to its own work methods without much regard for the impact on other participating functions. The impact goes beyond the normal considerations associated with functional interfaces: It deals with the internal lost time of each function. The objectives of each of the functional groups are obviously different, but focusing only on the needs of a single function works against the potential contribution of all toward building a successful and sustaining business. The functions shown in Fig. 2.2, the tripartite organization, will be the reference points for considering the work methods that generally waste time and result in longer cycle times.

Research

Scientists and engineers work at their own pace and generally take the step-by-step approach, each piece of information adding its bit to the storehouse of knowledge. There are obvious benefits to such an approach if knowledge and understanding are the ultimate objectives. But businesses devote a small percentage of their research allocation to fundamental or basic research, that is, research into the totally unknown. Most industrial research relates to products or processes and involves incremental improvements rather than breakthrough discoveries. Research scientists and engineers can conserve much time by taking the giant step when appropriate. The size of that step will differ, but not every experiment must be based on the results of the last one. Sometimes it can be beneficial to do the last experiment first and then go back to gain a better understanding of the underlying principles.

Development

Development departments, although more closely allied to manufacturing and marketing, also work according to their own methods. Development managers who insist on following every step in the process, regardless of the type of development, only waste the resources that could be used more effectively in some other manner. As an example, the introduction of computer-aided design (CAD) can be a blessing or a curse depending on how it is implemented. If every design must pass through the CAD system regardless of complexity or urgency, time will be wasted. If every development engineer must become proficient in CAD, resources will be wasted. Not everything is performed according to the three E's (effectiveness, efficiency, and the economic use of resources) on CAD. The design on the back of a napkin, developed over lunch, continues to serve its purpose, as does the visit to the machine shop or some other specialty function to mock up a model to demonstrate an idea or concept.

Managing engineering changes provides another example of wasted time: It involves time and money. The cost of engineering changes can equal the cost of the original design. Although the cost of preparing the engineering changes may be considered unimportant or inconsequential as a percentage of total cost, the time consumed in the activity and the time delays that arise as a result of changes can greatly affect the return on the original investment. Marketing a product that is on time and satisfies both the specifications and the customer's needs continues to be the best way to build a successful business. Engineering changes must be managed according to the three E's, and that includes the research and development stages.

Manufacturing

Manufacturing managers are usually schooled in the basics of industrial engineering. In turn, industrial engineering is guided by very definite methods for accomplishing certain tasks. That approach limits the degree of innovation that is acceptable to management. The objectives of manufacturing are to move the raw materials through the factory and into the warehouse or distribution center. The typical refrain of the manufacturing manager is: Don't bother me with any new technology. We haven't got the last system working, and that's been installed for over a year. Arguments against implementing new technology in the factory are real. Why does a manufacturing manager want more headaches? But it is important to look at the net effect of not implementing new technologies.

New manufacturing technology need not be a nightmare. There was no reason to fear the application of robotics or computer-integrated manufacturing (CIM). Ignorance of technology, as in other instances, creates fear. Practical implementation depends on the level of expertise of those involved and the up-front work that preceded its choice, design, construction, and installation. Design of new products and manufacturing processes must take into account the constraints under which the total system operates. Those constraints must be taken into account in the design stage of the product or process and not when production is ready to begin. The shakedown or the commissioning of a new process can be shortened by following a systematic checkout procedure that includes not only manufacturing but also research and development.

Marketing and sales

Marketing and sales also live in their own worlds with work methods that are often no longer acceptable. Marketing must describe its role and function in the business. What it includes is usually vague. Marketing has a role in new product planning that cannot be underestimated. It should be the major source of information about the needs of the marketplace. Too often, however, marketing is performed by sales people who have only a limited understanding of what marketing involves. Marketing goes far beyond maximizing the sales revenue. How marketing people work determines their level of effectiveness. There is a creative element in marketing that goes beyond advertising and extolling the benefits of the latest glitzy piece of promotional material.

The archives of most organizations contain marketing plans of many different descriptions. Those marketing plans usually present a biased perspective, and often the information comes from the wrong sources. The statistical information about the size of the markets and other pertinent information often comes from published industry data that most likely were never verified as to their applicability to and credibility in the specific situation. Determining the market needs involves going into the field and talking with customers or potential customers about their needs and also educating them about the possibilities of improving their operations. It is interesting to note that none, absolutely none, of the computer companies have managed to develop a new, comprehensive approach to secretarial services. All continue to sell more hardware and software, but without any simplification of or improvement in secretarial performance. The process has been totally ignored. Despite large investments in personal computers, printers, plotters, copy machines, and fax machines, the electric typewriter is still needed to type a single label efficiently. The marketing plans for computers and the rest focused on selling products rather than solving problems or improving productivity. They never looked at the system.

Physical distribution

Physical distribution seldom receives any attention until a valued customer suffers a mishap. Certainly many of the distribution operations have been automated in the past years, but customers continue to receive the wrong products and shipments often arrive after the scheduled date. Physical distribution does not function independently of marketing and sales. Three examples demonstrate the impact of marketing and sales on physical distribution:

- Order entry and processing

- Order size

- Special requests

Order entry and processing. Some may argue that order entry and processing do not involve marketing and sales. That depends on how an organization describes the function of marketing and sales. Satisfying the customer's needs involves moving the request for information or delivery of product through the system efficiently. A system that takes days to ship an item that is in stock is no longer acceptable. A distribution system that requires a physical check of the inventory before making a delivery commitment to a customer hardly meets current requirements.

Order size. It is only recently that businesses began looking at profit per customer and established minimum sizes of orders. Small orders were never profitable. They are usually justified by the expectation of business that seldom materializes. Unprofitable orders generally require additional time for processing, and they only consume time that could be used more effectively in pursuing other business opportunities.

Special rush orders. The special rush orders that every salesperson has generally create additional delays. That is not to suggest that special attention is never required. The number of times per day that such requests are made determines the index of dissatisfaction with distribution. The more disruptions that occur, the greater possibility that other customers will be dissatisfied with their deliveries.

Customer service

Customer service involves pre- as well as postsales activity. It begins with the attempts to sell the customer and ends when the product or service is delivered and the customer is satisfied with the performance. Recently, because of increased competition, businesses have developed special programs related to customer service, but in the real world not a great deal has changed. Executives may preach that customers are always right—in reality, they are not—but the follow-up actions fall short of satisfying the complaints. Undoubtedly, customers enable an organization to grow and prosper, and their many needs must be satisfied. Those needs go beyond just the product or service that was provided, however; they involve educating the customer. Sales and service people are the company so far as the customer is concerned. They will fulfill their customer responsibilities to the extent that they have been educated by the company in the importance the company places on customer satisfaction. The time spent in attempting to satisfy a disgruntled customer, regardless of the reason, is an added cost in time and money. In many cases, the customer complaints reach the office of the CEO. As the complaints make their way to the top of the organization and are resolved to the satisfaction of the customer, not only are costs added but valuable and expensive time is lost. Total customer satisfaction is achieved only when the time spent in dealing with complaints approaches zero.

Financial

Financial operations also must change some of their work methods to conserve time and eventually cycle time. Accountants educated in the fundamentals of technology and marketing are even more rare than the scientist or engineer with a business and accounting orientation. As an example, accountants, like the researchers, seldom have contact with the customer, the factory, or the distribution activities. They work from paper and often do not know whether the inventory is in the warehouse or whether raw material or finished product really exists and, if it does, what its value is.

There is no doubt that accounting records are essential. The question that must be answered is, How much information is required and to what accuracy? As an example, forecasts and budgets are at best estimates. They need not be calculated to the nearest penny. Yet budgetary line items such as $1,285,567.21 continue to be commonplace. The question is, Why? Why not a column headed with × 1,000 and the entry 1,286?

The paperwork to process an investment or authorize a sizable expenditure usually involves a specified number of signatures. Cycle time can be affected by delays created in accounting. If the technology and marketing functions delay involvement by accounting until a program is finalized, it is not uncommon to have 2- or 3-month delays in the accounting department. Effectively managed businesses would not allow more than one day for an authorization to clear the accounting department, regardless of the amount. If the accountants are brought into the process from the beginning, they need only affix their signatures and any pertinent financial information as the paperwork flows through the system. The financial analysis should have been completed long before the finalized paperwork arrives. The accounting function must be performed simultaneously with the technology- and marketing-related functions. Questions must be asked by finance during the project preparation period. That approach not only saves time and improves cycle time but allows the accountants to understand the different elements of the proposal and offer their input.

Human Resources

Human resource departments have expanded in recent years, and they should look not only at the time wasted within the department but also at the time wasted in the business as a result of their activities. Human resource specialists have tended to give the store away without receiving anything in return. There is no doubt that government regulations have complicated the activities, but not to the degree suggested. To let fear of community or government retribution delay for years the termination of a chemically dependent individual who, after repeated counseling at great cost to the organization continues the lifestyle, seems inconsistent with good business practice. Everyone can sympathize with the individual's problem, but what about the impact of such decisions on other employees? The decision to prolong the individual's employment sends a clear message: It doesn't matter what you do, the company won't terminate your employment. The hours of lost time accumulated by such human resource actions must be curtailed.

Robert Townsend [13] suggested that the whole personnel department should be fired and replaced with a one-person people department. Although such drastic action may not be required, human resource departments must recognize their contribution to lost time. The following activities are only a sample:

- Introducing a new human resource proposal without defining the benefit

- Soliciting input through employee surveys without adequate follow-up

- Emphasizing the "charm schools" instead of fundamental knowledge and understanding
- Misleading employees about potential for advancement, career planning, and so on
- Grasping at new behavioral fads without fully understanding the underlying principles or the research that led to the hypotheses
- Imposing themselves in the assignment and promotional processes of operating units

Patent and legal

Patent and legal people seldom appreciate working under time constraints, but the effective and efficient use of their time for matters related to technology is just as important as that of any other business function. Technology involves patents, contracts, and legal negotiations and thus requires the cooperation of both groups. There is no doubt that legal matters, like technology matters, must be analyzed systematically and in depth before decisions are made. But the depth of the analysis will vary from case to case. Not every i must be dotted, and not every t must be crossed. As with an engineering design, certain items will have been overlooked. It is always interesting to observe how people who were responsible for negotiating an agreement are almost eliminated from the adjudication process other than for their direct testimony. If the parties to the agreement were responsible for resolving the disputed issues, the time wasted in debating technicalities and prolonging the process would be reduced significantly. The legal minds cannot perceive all of the intervening events.

Public relations

Public relations departments seldom include individuals capable of communicating matters related to technology. As a result, it often becomes necessary to assign engineers and scientists to spend time educating them and subsequently reviewing their press releases. High levels of technological expertise are not expected in public relations departments, but some moderate level of understanding would conserve the valuable time of costly engineers and scientists. The advantages of new technologies and new products tend to be exaggerated. Technology announcements tend to skew the thinking processes of rational people. Optimism and speculation take precedence over facing up to the realities. The public relations releases about expert systems, artificial intelligence (AI), CIM, robotics, automation, and so on have never agreed with the realities of application. Executives and managers began to believe their own public relations hype. Romancing of new ideas, concepts, and technologies is essential, but at some point before major investments in time and money are made, a dose of realism must replace overly optimistic speculation. All these activities consume time and extend cycle time.

Purchasing

Purchasing departments and the primary technology functions of research, development, and manufacturing determine manufacturing costs. That applies equally to products and services. But purchasing often contributes to lost time and long cycle times. Purchasing, like the other business functions, often deteriorates into a bureaucratic fiefdom. If it is conceived as the final arbiter and decision maker in all purchases and deals with the lowest bidder as a matter of practice, the final outcome is not only added cost but delay in shipping. Few purchasing de-

- Audio visual services
- Benefits administration, pension, and so on
- Communication services: telephone, voice and electronic mail, audio and video conferencing, telecommunications, and fax
- Company store
- Custodial services
- Data processing
- Economic studies
- Education and training
- Environmental requirements
- Fire and emergency evacuation
- Food services: cafeterias, dining rooms, vending machines
- Library services: books, periodicals, reference, circulating, vendor information, market studies, subscription orders, and databank searches for all of the business functions
- Logistics
- Mail service within and outside the organization
- Maintenance of office equipment
- Medical: health and hygiene
- Office maintenance and rearrangement
- Supplies and office equipment
- Toxicology services
- Transportation: company cars, aircraft, and airport limousines
- Travel services and analysis
- Waste and scrap disposal

Figure 3.3 Common general administrative activities that, although taken for granted, affect aspects of managing cycle time.

partments look at total costs or system costs. "Cost" includes more than the cost, handling, and delivery of the specific item. Out-of-specification material can increase the manufacturing cost many times over, as well as create delays in production scheduling. Purchasing departments that deal in technology matters, whether related to raw materials, components, subassemblies, equipment, hardware, or software, contribute not only to the cost of lost time but also to extended cycle time.

General Administration

General administration comprises all the activities without which an organization cannot operate. The services provided are like water and other elements of a society's infrastructure: They are expected to be available and are taken for granted. The activities or functions vary considerably from one organization to another; Fig. 3.3 lists the common ones. The performance of the activities is seldom considered until there is a major problem that can be traced to their nonperformance. General administration also generates lost time that affects the technology functions. Information affects all of the activities of the technical functions. It is not limited to what is produced by the information management systems and is related to costs and financial results. It includes the library resources such as books, periodicals, and journals. It also includes the competitive information about technology, marketing, and management. The technology data banks must be available to scientists and engineers to prevent reinvention of the wheel. The effectiveness with which those activities are accomplished determines the lost time and the eventual effect on cycle time.

Time Frittered Away by People

Although 100 percent productive utilization of time can never be achieved, every individual could make much better use of time. Some lost time occurs because of overly restrictive policies and procedures, some because of delays in decisions, but most because of a lack of business discipline that begins with top management. Discipline does not mean conformance, strictness, punishment, or being tough. It involves a balance of order and freedom to act. It establishes limits of acceptable performance with an emphasis on raising performance standards continually as new knowledge, experience, and management tools are acquired. Discipline involves developing an environment in which people will do what they agreed to do without excuses. It involves making decisions in a timely manner. It begins in the CEO's office. If a lack of discipline exists at the top, there is no reason to expect it at lower levels.

Time is frittered away by people who work without any direction and do not know where they are going or what objectives they should be working toward. The responsibility for this frittered-away time can only be assigned to managers and their superiors. This lost time involves individuals and groups and the time frittered away in meetings. Time frittered away is not only lost by not doing something; more often it is lost by working at something very diligently and with great dedication. As an example, consider the case of time frittered away by working very hard: Organizations that are involved in global operations work in many different currencies. The situation to be described involves an international engineering operations manager who had an employee convert the capital forecasts of a large number of subsidiaries into dollars every month or whenever currency exchange rates fluctuated. It was not just a matter of converting the totals, the bottom line; every item listed on the forecast had to be converted. Thousands of projects were involved. There were no reasons for such extensive conversions. What should have taken several hours kept the individual and eventually an assistant occupied full time. Such activities must be eliminated. They serve no useful purpose.

Technical staff groups, whether related to research, development, or manufacturing, that function as consultants to operating divisions often consume time and provide no added value. In most cases the operating divisions do not pay directly for the services; the costs are absorbed and then distributed according to some formula among the operating units. There is no doubt that special expertise is essential and cannot reside in every operating unit. It could not be justified. But how much of that expertise and consulting is actually used poses some major problems. My research has shown that internal specialists or consultants add value to a project only when they are active participants. They have a role to play in the success of the project, but they must contribute more than advice. If operating units were paying directly for the use of specialists or internal consultants, not only in technology but in any discipline, more productive use could be made of those people. In the process the amount of lost time would decrease.

Every organization has its own origins of lost time in activities that serve no useful purpose. Much technical reporting falls in that category. If needed, it should be brief, concise, and devoid of any information that serves no purpose. For some reason, managers feel more comfortable if they have the report in their hands. Why not just take a hands-on approach and find out what is really going on in the organization? One could argue that such an approach may be acceptable only with first- or second-line managers, and that may be true. If it is, why not structure the written reports in such a way that unresolved problems, delays in schedule, or significant design changes are explained rather than smothered by excessive verbiage, often intentional, that conceals the true status.

Besides the time frittered away as a result of managerial inaction, much time is lost by the idea originators and their peers. In recent years, innovation and intrapreneurship have received attention comparable with that given quality, empowerment, and culture. Much of the innovation has not materialized, and companies like Eastman Kodak [14] that have attempted to introduce an intrapreneurial culture have not been very successful.

Much of the time lost by the creative engineers and scientists and marketers arises from a lack of knowledge of the system. The creative people do not know how to function in the system. Granted that some managers may be incompetent, inattentive to new ideas, or in fact be disinterested in pursuing new ideas. If no practical way can be found to circumvent such behavior, there is only one long-term alternative: Find someone within the company who is responsive. If that is not possible, look for an organization with a more appropriate environment. That may be drastic advice, but if there are no possibilities of changing a manager's attitude, why ruin your own career? Many employees find themselves in just such a predicament. This example, which fits many employees costs companies large amounts of lost time and eventually costs the individuals productive and satisfying careers.

Time lost by idea originators [15] or creative individuals arises from not understanding the purposes, objectives, and strategies of an organization. The value of an idea is a matter of perception. What is creative and useful to one person may be mundane and impractical to another. Logic and rational decision making do not always guide business directions. Eliminating this lost time requires some business-related knowledge by the originators. Figure 3.4 lists the business-related knowledge that can help idea generators and innovators speed up their programs.

Peers can be a blessing or a curse. They can be supportive, provide their expertise, critique your approaches, help find a sponsor, and become part of the network that is essential for accomplishing any activity that involves different professional disciplines and the involvement of other business functions. That is the positive side. But beware of the peer who is not fully occupied or for some reason prefers chitchat, the grapevine, scuttlebutt, the rumor mill, or just plain gossip to settling down and completing a project on schedule. Such peers occur at all levels in the organization from top to bottom. They are part of management, the professional staff, the secretarial staff, and the production staff. They waste not only their own time but the time of others. The important issue is the wasted time of others.

1. Understanding the organization and its management—involves more than knowledge of the organization chart.

2. Finding a sponsor to support the idea. Even simple experiments require funding, either authorized or bootlegged.

3. Reducing the concept to the essentials. This takes time and hard work to fully understand the principles and develop a level of confidence.

4. Finding a way to make people comfortable with the idea. Every person operates on a private time schedule and requires different amounts of information.

5. Considering the risks and uncertainties. Any idea that is worth pursuing involves some level of risk and uncertainty. Understand those risks and uncertainties upfront and find ways to reduce them.

6. Soliciting management support. A sponsor is required as noted, but so is the support of other professionals in research, development, manufacturing, and marketing.

7. Knowing competitive approaches—before projects are accepted or funded the competitive issues must be addressed—do not wait to be asked.

8. Selling product or business concepts, not technology. Management accepts your technical competence; tell it what that will accomplish for the business.

Figure 3.4 Business-related knowledge that can help idea generators and innovators move an idea or concept through the system.

Indirect Influence

Thus far, this chapter has focused attention on *direct lost time:* time lost by the system, management, business functions, and people. However, time is indirectly lost by the lack of interaction among the generators of lost time. Little attention is given to the lost time imposed on the organization by the total system in which the business operates. That time is lost or used inappropriately but nothing can be done about it is commonly accepted as fact. The result is an attitude within the organization that reducing lost time is not a high-priority issue, so there is no reason to be concerned about the ineffective use of time. But system lost time affects the whole organization, and the amount of time lost increases as the complacent attitude flows through the organization.

Management, through its actions, establishes the attitudes toward time. Those attitudes will either focus the business unit on the impact of lost time or ignore it. Purposes, objectives, and strategies of a business unit lie on a continuum. Objectives cannot be developed without well-described purposes (missions), and strategies can be developed only after the purposes and objectives are clearly delineated and understood. Thus, a decision by a manager in one function can adversely affect the time lost in other functions. A decision to go into production on a product when prototype performance has not been verified will not only create unnecessary problems in manufacturing, with immense amounts of added time, but may lead to the loss of valuable customers. Managers must look at the net effect of their decisions on the business rather than on the benefits that could be ascribed to a single functional unit.

The work methods in one functional group can affect the time lost in other functional groups. The cost of those who stand and wait for whatever reason not only lowers productivity but eventually reduces performance standards. Consider the four functions of research, development, manufacturing, and marketing. When does research involvement in a new product introduction end? Not until the product has been commercially accepted and operates according to specifications. Research or development may consider that a responsibility of customer service, but that is so only if the customer service people have the talent to resolve the problem and satisfy the customer. There are some encouraging reports about the use of cross-functional and multidisciplinary teams, but their net long-term effectiveness and efficiency has yet to be determined.

Without any doubt, people's actions are influenced indirectly by others outside their immediate workstations. The same people have the opportunity and the means for indirectly influencing the actions of others outside their immediate work places. That indirect influence can be either positive or negative. It conserves valuable time, or it contributes to lost time. Although some of the influence can be beneficial and supportive of the objectives of the business unit, much of it can degenerate into chitchat with no specific purpose in mind. Managers must be cognizant of the lost time that is generated by activities outside their immediate spheres of influence.

External Influence

External forces and activities influence the time lost by the system, management, functional groups, and people. Those four origins of lost time make their contributions in all organizations from the smallest to the global corporations. The complete business organization and each of its many subdivisions are affected by external decisions, forces, and activities. The major special-interest groups include customers, suppliers, investors, all levels of government, the local communities, the condition of the national economy, and, more recently, global events.

Customers

It is true that customers are the vital force of the business, but some customers can be troublesome and tiresome even though they may be excellent from a financial point of view. They demand more service than required and continual hand-holding.

Suppliers

Suppliers of all types and descriptions also cover the spectrum from the reliable to the unreliable. Why organizations tolerate unreliable suppliers is difficult to understand. Perhaps it is the result of a social phenomenon: Accept mediocre performance and pay for it.

Investors

Private investors occasionally consume much of management's time, but none to the extent of the investment houses. There can be no rational reason why organizations submit to such procedures. What is interesting to note, though, is the authority with which representatives of the investment community speak. It would be difficult for anyone who knows the inner workings of a company to accept their reasons for investing.

Government

Government at all levels imposes large doses of lost time on the business community. That is not to suggest that the government is not entitled to reports and inquiries. However, the process has become unmanageable and self-defeating to the extent that the information provided by industry does not appear to serve any useful purpose, except for attacking some specific organization. The industrial committees that the government periodically organizes to improve its efficiency and effectiveness make recommendations that are never implemented. Their reports are accepted with typical ceremony, but nothing changes. The J. Peter Grace report [16] to President Reagan on cost control in the federal government involved 36 task forces, 161 corporate executives, 2000 volunteers, and provided 2478 separate, distinct, and specific cost-cutting and revenue-enhancing recommendations. It is difficult to find any recommendations that have been implemented. The David Packard Commission report, *A Quest for Excellence* [17], on Defense Management, was received with enthusiasm, but the bureaucracy failed in the implementation. These are examples of time wasted by government and by the many volunteers in industry.

Communities

Businesses cannot ignore the nongovernmental aspects in the communities in which they function. As a rule, business is looked on as the provider of jobs. Although most major business organizations try to be socially responsible, within the limits of their resources, more always seems to be demanded of them. There is a practical limit to the resources that can be allocated to nonproductive activities. The needs of the stakeholders must be balanced.

National economy

National economic conditions also generate lost time. That is primarily true of a static or declining economy. New programs are started without much long-term thought. Decisions that are made may be counterproductive in the long term but provide a short-term benefit. Management pushes the panic button, loses its focus, and tries to pass the blame for not meeting expectations on to others. Good executive performance involves meeting business objectives even under adverse economic conditions or, at a minimum, anticipate the adverse conditions.

Global events

There is no doubt that global events consume time that is generally nonproductive. In recent years, conversation alone has cost many hours of speculation on the outcome of successive world-shaking events. But as companies consider investing in the eastern block countries, the lessons learned from investing in other foreign countries seems to have escaped executive management. Those treks to explore business opportunities, in the absence of any understanding of the cultures and the real economic conditions that exist, will not lead to successful joint ventures.

All these external powers and influences consume time that could be spent on business growth. The amount of time spent as a result of external forces is looked upon as the necessary cost of doing business. They should not be accepted as such. Time lost by such activities can be substantially reduced without negatively affecting any of the constituents. The time that is saved can be allocated to more appropriate activities that promise to improve the performance or expand the business.

The Cost-of-Lost-Time Report

Attempting to determine the origins of lost time should not become another new major business program. It is part of management's responsibility inasmuch as where time is allocated determines the quality of business performance. Searching for the origins of lost time goes far beyond the traditional approaches to managing time. It goes beyond making schedules and extensive to-do lists, eliminating extraneous phone calls, curtailing long lunches, reducing the time lost by waiting, and so on. Those steps are important but insufficient. Time management courses provide little long-term benefit. Such courses may attempt to convey a methodology for organizing time, but more is required.

In a *Fortune* article [18], Jack Gordon, the editor of *Training,* was quoted as saying:

> The truth is that if you are already kind of logical, linear, and organized, a time management course will make you more logical, linear and organized. If you're not—if you don't want to organize your life into A, B, and C priorities—then you won't.

Managing time in relation to timing and cycle time must go beyond a linear approach. It goes beyond adding new gadgetry such as electronic mail, computerized telephone directories, and fax machines. It requires discipline in focusing resources on the stated purposes, objectives, and strategies of the organization or business unit. It involves determining what is worth doing. Putting priorities on nonessential activities that do not add value serves little purpose. Executives may set their priorities, but they also must recognize what they are delegating. They may consider themselves efficient users of the time resource, but how much consideration have they given to what they delegate or pass down through the organization? How much wasted time are they generating?

A cost-of-lost-time report may be one way to call attention to what that lost time really costs an organization. As was stated earlier, it probably exceeds the net profit of even the well-managed organizations. That should not be an additional task for managers, since it is one that is essential and cannot be avoided. Unless organizations identify the origins of lost time, they have no opportunity to decrease it and subsequently eliminate it. Developing a cost-of-lost-time report does not require hiring additional people to do extensive surveys. It does not involve controlling time. It does involve appraising or auditing how time is allocated and spent in meeting the organization's purposes, objectives, and strategies. The report would include a description of the source, the hours lost per some period of time, the cost per hour, and the total cost with an accuracy of plus or minus 10 percent.

Figure 3.5 lists the practices that are responsible for lost time, and it applies to each of the functions in the tripartite organization. It provides some guidance as managers begin to investigate the origins of lost time. The lost-time report is not intended as a monthly report; it is an instrument for periodically evaluating the effectiveness of the use of time for the benefit of the business.

Summary

This chapter has presented the origins of lost time that far exceeds any time that might be saved by the use of mechanistic tools. The elemen-

1. Ignoring complacency and lack of motivation at all levels of the organization: executives, managers, professionals, and all other employees.
2. Communicating at all levels of the organization so that the message gets through rather than the noise—the real intent, not the hidden meaning.
3. Deviating from or not communicating changes in purposes, objectives, and strategies.
4. Losing touch with reality and not knowing when romancing an idea must be replaced by realistic decisions based on some probability of success.
5. Using single issues, fads, quick fixes, and slogans instead of integrated approaches to resolving performance and productivity problems.
6. Promoting policies that consume time but provide no measurable benefit.
7. Insisting on adherence to restrictive and rigid procedures that in effect curtail the creative and innovative spirit that executives hope to engender.
8. Delaying decisions at the executive level and then expecting the operating groups to shorten their time for accomplishment.
9. Allowing the bureaucracy to become inefficient and detract from expected business performance.
10. Developing a welfare mentality among the employees, which includes the professional staff and executives and managers to varying degrees.
11. Attempting to engage in activities, businesses, technologies, and markets without adequate knowledge and competence.
12. Neglecting the allocation of all business resources: focusing on one resource rather than the whole system.
13. Managing the information transfer system as a confidential data bank rather than as an information resource easily accessible by those who have a need to know.
14. Providing coordinating functions that only consume time and add confusion to the process.
15. Changing priorities too often without sufficient information to the program or project participants.
16. Accepting illiteracy at all levels in the organization on matters related to business, technology, markets, finance, and human resources.
17. Omitting the up-front thinking before making decisions of any kind. Up-front thinking does not take additional time; it conserves time.
18. Misunderstanding the role of leadership in business performance—the need for proactive participation without interference.
19. Neglecting the importance of business discipline—not being tough, but fostering a mental and attitudinal discipline.
20. Promoting the game playing, such as contests, that focus attention on maximizing the results of some single activity without consideration of the associated increased costs in other areas.
21. Neglecting the development of advanced work methods that are required as new and advanced technologies are introduced into the business process.
22. Condoning the inefficiencies that result from functional dynasties, which exacerbate the interface problems.

Figure 3.5 Practices that are responsible for lost time.

23. Insisting on reports or report formats that bloat the archives without any rational purpose—the generalized reports for everyone's attention that seldom if ever are read or acted upon.
24. Working below the capability level for extended periods of time without a realization of the total impact on the individual and the other personnel.
25. Overlooking or often rewarding supervisors who are not capable of making judgments related to acceptable and nonacceptable levels of performance.
26. Excluding knowledge of how the organizational system functions from the educational process. Employees must know how the system works.
27. Participating in the many extracurricular community functions that provide personal ego trips but do not benefit the organization.
28. Going on business trips that could be covered more appropriately by phone calls or visits by a local representative. Personal contact with clients of all types is essential, but it can be accomplished efficiently and effectively.
29. Attending meetings that supply you with information you do not need or at which you make no contribution to the results.

Figure 3.5 *(Continued)* Practices that are responsible for lost time.

tary tools are important because they tend to develop a discipline in the use of time: They are only part of the answer. The major issues that managers must address relate to the value of dedicating time to certain specific activities. That does not imply that every effort will be successful. It does imply that thinking will take place before action. A researcher may fail many times while plowing new scientific ground, but the methodology and the scientific approach cannot be left to chance. Not every new product development will meet expectations. Not every marketing plan will be successful. Not every management decision will provide the expected results. But in each case, it is expected that those involved will practice their profession or trade with an understanding that *time* is a limited resource.

References

1. K. Rebello and J. Schneidawind, "Seeing Red over IBM's Blues," *USA Today,* May 31, 1991, pp. 9B–10B.
2. J. A. Byrne, D. A. Depke, et al., "What's Ailing Big Blue?" *Business Week,* June 17, 1991, pp. 24–32.
3. P. B. Carroll, "IBM Plans $3 Billion Charge and about 20,000 Job Cuts," *The Wall Street Journal,* Nov. 27, 1991, p. A3.
4. P. B. Carroll, "How an IBM Attempt to Regain PC Lead Has Slid Into Trouble," *The Wall Street Journal,* Dec. 2, 1991, p. 1A.
5. T. J. Peters and R. H. Waterman, *In Search of Excellence,* Warner Books/Harper & Row, New York, 1982.
6. S. P. Sherman, "Digital's Daring Comeback Plan," *Fortune,* Jan. 14, 1991, pp 100–103.
7. *Application Guidelines for The Malcolm Baldridge National Quality Award,* the United States Department of Commerce and the National Institute of Standards and Technology, 1991.
8. "The Ecstasy and the Agony," *Business Week,* Oct. 21, 1991, p. 40.

9. Y. K. Shetty, "Product Quality and Competitive Strategy," *Business Horizons,* May-June 1987, pp. 46–52.
10. G. Pinchot III, *Intrapreneuring,* Harper & Row, New York, 1985, pp. 137–142.
11. J. A. Byrne, D. A. Depke, et al., "What's Ailing Big Blue?" *Business Week,* June 17, 1991, pp. 28–29.
12. V. Ramanujam and N. Venkatraman, "Planning and Performance: A New Look at an Old Question," *Business Horizons,* May-June 1987, pp. 19–25.
13. R. Townsend, *Further Up the Organization,* Alfred A. Knopf, New York, 1984, pp. 172–173.
14. J. S. Hirsch, "At Giant Kodak, Intrapreneurs Lose Foothold," *The Wall Street Journal,* Aug. 17, 1990, pp. B1, B8.
15. G. Gaynor, *Achieving the Competitive Edge through Integrated Technology Management,* McGraw-Hill, New York, 1991, pp. 236–247.
16. J. Peter Grace, Executive Committee Chairman, "President's Private Sector Survey on Cost Control," *Document No. S/N 003-000-00616-6,* Superintendent of Documents, Washington, D.C., Jan. 15, 1984.
17. David Packard, Chairman, Blue Ribbon Commission on Defense Management, *A Quest for Excellence,* Superintendent of Documents, Washington, D.C., June 1986.
18. E. Calonius, "How Top Managers Manage Their Time," *Fortune,* June 4, 1990, pp. 250–262.

From Macro to Micro

*To look at cycle time from a macro perspective is to lose
operational significance. As with so many management
concepts, it is necessary to differentiate and delineate the many
activities and their differences and shades of meaning. As an
example, the introduction of strategic planning cost
organizations vast amounts of money with little if any real
benefit, that is, benefit when measured against the
expenditures of time and money and the sales lost by diverting
scarce resources. The strategy planning gurus sold corporate
executives the concepts without the executives understanding
what they were buying. The executives accepted as fact that the
concept called strategic planning would lead them leisurely
and successfully through uncharted paths to a more
competitive organization.*

*Part 1 considered the limitations of describing cycle time in
the limited and restricted terms of concurrent or simultaneous
engineering and improving communications between
marketing and research, and so on. It also included an in-
depth description of relations among business unit
organization and the three components of cycle time and
introduced the system cycle time model. Since time plays such
an important role in optimizing cycle time, examples were
presented to take the reader beyond the simplistic approaches
to time management. The objective was to provide the reader
with some insight as to where the time wasters originate and
to demonstrate the need for taking the total business system
into consideration.*

*Part 2 focuses attention on four specific types of cycle time
ranging from the macro to the micro. They include business*

system cycle time, product cycle time, project cycle time, and time to decision. Emphasizing any one of these four types will not provide a significant competitive advantage. Managing all four types and understanding their interrelations and their interdependence can make a significant difference in total business performance. This classification, chosen by the author, is based on personal experience, observation over many years of how industry mismanages cycle time, and research into the multitude of factors that affect it.

Chapter 4 directs attention to business system cycle time. It expands on the influences of the business system that determine the time to market. How an organization describes system cycle time makes a major difference. Is system cycle time the time from concept to successful commercialization? Is it the time from authorization to successful commercialization? Is it the time from order to delivery? Is it the time from receipt of an order until the check clears the bank? Is it the time from order until the customer receives what has been ordered, including meeting all the operational specifications? Does it go beyond those descriptions? Managing cycle time begins with an understanding of the issues that affect it.

Product and process cycle time, the subject of Chap. 5, draws attention to three specific areas: (1) the factors that affect the length of time it takes to bring the product to market, (2) the time during which the product will meet market requirements, and (3) the time for obsoleting products. Optimizing product cycle time requires more than a simple concern for calendar time. Additional costs could be involved. It must also be based on the value of that product from a technology and marketing viewpoint and the product's contribution to profitability and relation to meeting the purposes and objectives of the organization.

Effective project management determines to a great extent the success of new product introductions. Chapter 6 raises the issues related to size, scope, complexity, number of disciplines, and business functions. Project cycle time receives the greatest attention from business executives. That is the level at which executives can touch and feel, the level at which they are comfortable. The chapter deals with the management problems, inconsistencies, disruptions, and discontinuities that force organizations to reallocate resources and, in the process, extend cycle time.

The consequences of time to decision and its effect on cycle time are the subject of Chap. 7. It is difficult to determine the actual time lost or the time wasted by the lack of timely decisions. Whether those decisions are delayed by the board of directors or any intervening level of management, all contribute a share to this lost-time category. Time to decision has been isolated because of its importance and its high cost. The time lost by a delayed decision increases geometrically as its effects flow through every level of the organization. The time lost or wasted by extended time to decision involves every individual in the action stream. Time to decision represents a critical issue in optimizing cycle time at the business system, product, or project level.

4

Business System Cycle Time

There is no lack of discussion or opinions about the role of competition in business performance. The press, business publications of all types, and academic journals continually lament the decline of competition. Some are now attempting to demonstrate that U.S. industry has reached bottom and is on the way up. Just exactly where an industry or a specific business lies on this upward trend is not too important. The fact remains that U.S. industry must plot an arduous path to retain its industrial dominance. The time has come for all businesses to begin a reality check that lays open their business operations and practices. Such a reality check would help determine if what is being done and how it is being done meet the requirements of the present business milieu. Such self-examination will provide the information from which to launch tomorrow's management approaches in creating more productive enterprises. The reality check, or self-examination, will reveal that businesses take the piecemeal approach to managing and disregard the benefits of managing the system. At some point, the rhetoric must match the reality.

Exploiting business system cycle time as a means of gaining competitive advantage depends on creating an environment for continuous innovation, managing technology in its broadest perspective, balancing the quantitative and the qualitative information in the decision processes, and optimizing those ten resources by following the fundamentals of good business practice: following the basics.

Business system cycle time is influenced by many factors. It is not a single issue that can be dealt with in isolation. This chapter considers the critical issues associated with business system cycle time:

- Scope of system cycle time

- System cycle time model
- Limits of the business system
- Managing the system
- Strategic and operational issues

Scope of System Cycle Time

The classifications of system, product, project, and time-to-decision cycle time have been introduced not to complicate the concepts of cycle time, but to put the concepts in the context of the total business. The classifications force organizations to recognize that consideration of cycle time must go beyond such limited approaches as simultaneous engineering. How the business system is described determines the scope and the limits of the opportunities for improving performance. Is it limited to some of the possibilities suggested in Figs. 2.4 and 2.6, or does it include the time from:

- Customer order to delivery
- Customer order to check clearance by the bank
- Customer order to full benefit received by the customer
- Idea disclosure to successful commercialization
- Concept definition to successful commercialization
- Demonstration of feasibility to successful commercialization
- Authorization to successful commercialization

How an organization describes the outer limits of cycle time determines how it thinks, that is, how it approaches the daily activities required to sustain its business. At a minimum it must include the time from receipt of the customer order until full benefit is received by the customer.

Figures 2.4 and 2.6 illustrated some cycle time starting and ending points. Points *A* and *B* are the easiest to describe, since organizations work with them continually. Figure 4.1 combines Figs. 2.4 and 2.6 and extends the time from *A* to *B* to include the time from *X* to *Y*. The lines from *A* to *X* and *B* to *Y* are broken to indicate that, in reality, *A* to *X* and *B* to *Y* are unlimited and could far exceed the time from *A* to *B*. The examples shown at points *A* and *B* are some of the more dominant limitations. The list is not complete and must be adjusted to meet specific organizational needs. The starting and ending points can be combined in any manner according to management's perception of cycle time. Cycle time could be considered from *scouting and*

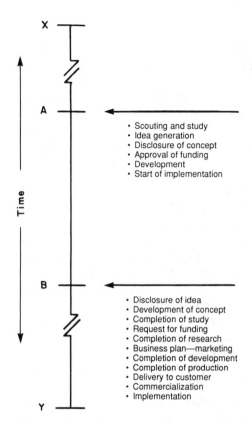

X

A

Time

B

Y

- Scouting and study
- Idea generation
- Disclosure of concept
- Approval of funding
- Development
- Start of implementation

- Disclosure of idea
- Development of concept
- Completion of study
- Request for funding
- Completion of research
- Business plan—marketing
- Completion of development
- Completion of production
- Delivery to customer
- Commercialization
- Implementation

Figure 4.1 The limits of cycle time.

study at point *A* to *commercialization* at point *B*. It could also be limited from *idea generation* at point *A* to *disclosure of idea* at point *B*. Organizations consider system cycle time differently. As an example, Motorola focuses on reducing the time between when an order is placed and when it is delivered [1]. Westinghouse, on the other hand, considers the time between conception of an idea and introduction of the new product.

The often forgotten extensions in cycle time occur from *A* to *X* and *B* to *Y*. The infrastructure of the business system, as illustrated by the three-dimensional model in Fig. 2.7, governs to a great extent just what occurs during the period from *X* to *Y* in Fig. 4.1.

The seven elements of the infrastructure have been researched many times without any definitive conclusions. What has been missing, though, is sufficient research in relation to their interaction. Those elements must be integrated into the evaluation of the system's lost time. They are the reasons for lost time that occur between points *X* and *Y* in Fig. 4.1.

Purposes, objectives, and strategies

There is little doubt that cycle time is extended if the purposes, objectives, and strategies of an organization or its subunits are not clearly described. That is self-evident fact. The same statement applies to any subunit of an organization, whether a function, project, or other organizational configuration. The purposes of an organization must be developed before any objectives or strategies can be described. A change in any one of the three requirements will affect the others. That applies to technology, markets, and the financial considerations of the organization.

There are those who flippantly state that the only purpose of the organization is to make money. That is an obvious but simplistic conclusion about any profit-making organization. Making a profit is essential, but profit results from fulfilling the purposes of the business. If the only purpose of a business were to make money, then why take the risk of investing in the resources normally associated with a product-driven organization? Introducing new products is a high-risk business, at least in the manner in which it is currently practiced.

If an organization cannot determine in which business or businesses it wishes to participate, and then develop its objectives and the follow-up strategies to accomplish those objectives, valuable resources will be squandered. What is the cost in lost time and the extension of cycle time when a decision is delayed for 5 years to make a major capital investment? Then when the decision to proceed is finally made, management expects the system to be designed, fabricated, installed, and in operation in 18 months and usually at additional cost. Actions such as those are usually rationalized in many different ways. The net effect is not only to increase lost time in the technology functions but also to affect the timing of product introductions and the subsequent manufacturing cost. A McKinsey & Co. study [2] showed that a product 6 months late to market misses out on one-third of the potential profit over the lifetime of the product. One can argue with such statistics, but whether the fraction is one-third, one-fourth, or one-tenth is of little importance: The simple fact is that there is a loss in profitability.

Organizational structure

Organizational structure accounts for some of the time lost from X to Y in Fig. 4.1. Whether the structure is complex or simple depends on the position from which it is viewed. There is no need to become particularly concerned about structure, since experience demonstrates that, with the appropriate combination of people with diverse talents, structure is essentially irrelevant. Without adequate talent and accommodation among the participants, no structure will make a great deal of difference in performance. The time lost from organizational structure comes primarily from three activities: (1) paperwork moved from one place to another by people who lack either the necessary knowledge or the motivation to act on it, (2) limiting the opportunities for independent decision making, and (3) substituting paper and more paper and more terminals for face-to-face communication. As an example, the approval of any expenditure can be delayed for months simply by having the paperwork sit on a desk because a signatory is not available and there is no procedure for avoiding such a delay. Someone in the approval chain can easily delay a program for many weeks. It makes little difference whether that approval involves technology, advertising, production schedules, or people assignment. A further complication arises when the "casa madre" (the home office) adopts the attitude that it is the source of all knowledge.

The guiding principles and the policies and practices

Inasmuch as the guiding principles and the policies and practices establish the culture of an organization, they affect the time from point X to point Y. They determine what management values and what it rewards.They establish the limits of freedom, flexibility, levels of conformity, acceptance of change, and limits of creativity and innovation. All those elements affect the system cycle time. Individuals who are rebuffed by their managers in submitting a new idea or concept may choose to adopt an attitude that the company is really not serious about listening to new ideas. Others who are told to stay in their own court or their own sandbox will not use their mental capacities in such a way as to optimize the output of the organization. Individuals who are convinced that management does not level with them on issues that involve them or about which they are concerned will mentally retire at age 30 but remain on the payroll until age 65.

Management attitude and astuteness

Lost time can also be attributed to management attitude and management astuteness. A CEO or other senior executive who keeps one foot in the office and the other foot on the golf course, in the boardroom of another organization, in a presidential committee, or in one of a multiplicity of local community committees passes the wrong message to the organization. Excessive participation does not demonstrate social responsibility if it occurs at the expense of the stakeholders. The question that must be asked is what effect these ego trips have on business performance. As a group, managers affect the organization in the same way. If the major source of satisfaction for executives and managers comes from participating in extracurricular activities and it is recognized as such by the participants, an attitude of why worry soon infests the organization. The marginal or non-value-adding activities negatively affect the three elements of cycle time: total time, timing, and cycle duration.

Management astuteness can lead an organization toward very positive results. Astuteness involves many different characteristics. It includes such qualities as

- Breadth of knowledge
- Understanding of interlocking relationships
- Discriminating between nuances of meaning
- Contributing to the creative and innovative needs of the organization
- Being practical, perceptive, insightful, and logical but not too logical
- Being mentally alert and discerning
- Being comprehensive but within realistic limitations
- Being just, believable, and creditable

How does that astuteness affect cycle time? If the characteristics are embodied in the executives and managers, they determine the work ethic of the organization, but work ethic in the sense of how people work rather than the traditional meaning of keeping their noses to the grindstone. People do not follow like sheep: They think, and their thinking defines the organization.

System Cycle Time Model

Design of a business system, like the design of a product, requires analysis and synthesis of data, information, and knowledge and subsequent reduction to the simplest form. However, the simplest may be complex. The result represents the simplest available yet adequate input at the time. Too often, product design problems are accepted as unavoidable because of lack of knowledge of the possible or the availability of totally different technologies. Elegant designs may provide a challenge to the designer, but they must also satisfy the other criteria that determine business success. The same approach must be applied to management. Methodological elegance in the design of business systems is not the goal; the goal is organizational structures that allow the business to conduct its affairs with the minimum number of conflicting interfaces as it responds to customers' requests.

Figure 4.2 presents an approach to considering the system cycle time that begins with an idea and develops into a commercially successful

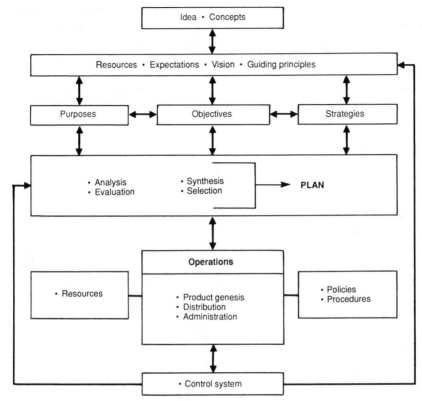

Figure 4.2 The cycle from idea or concept to commercialization.

product or service. There are many ways in which to describe that process. It is necessary to understand that control of any process—manufacturing or management—depends on feedback. However, feedback involves more than communication of some deviation. It provides a benefit only if it is timely, understood, and acted on. Some change must take place as a result of the feedback.

Figure 4.2 begins with ideas and concepts that are generally inadequate as presented. They are the seeds and not the harvest. Those initial ideas and concepts seldom embrace the resources, expectations, vision, and guiding principles of the business. Every organization, of necessity, limits its entrance to other markets by those four conditions. As an example, an idea for a new type of supersonic aircraft would not receive much consideration from a manufacturer of semiconductors. Likewise, an aircraft manufacturer would most likely not be interested in investing resources in women's apparel. Ideas and concepts add value only when they are achievable within the limitations of the business resources. To state it another way: If the business is involved in growing and selling apples of all types and descriptions, bringing in orders for oranges does not provide any added value. As a matter of fact, such unrelated effort consumes time that should be directed toward the purposes and objectives of the organization. It is a waste of the *time* resource. This example brings the question back to the limits of the process. The process will not respond to oranges; it is designed for all types of apples.

Ideas and concepts must also fit the purposes, objectives, and strategies of the business. The three are interdependent. They must be stated clearly and concisely at every level in the organization. In addition, they must be screened after an analysis, evaluation, and synthesis of the available information. The selection process depends on the screening criteria established by the organization. That initial screening information originates at the top of the organizational pyramid. It reveals the areas or domains of business interest. The screening process can take many different forms, but it begins with a clear understanding of the purposes of the organization and culminates in a clearly delineated and holistic operational plan.

The operations section includes the three entities of the tripartite organization structure illustrated in Fig. 2.2. Operations determines what actions will take place in transforming an idea or concept into a commercially viable product or service. Resources, which include the ten previously stated, are cited again and attached to operations, since now they must be realistically evaluated as to their availability. They will determine the probability of success. Policies and procedures are attached because they guide the organization in meeting its objectives. They determine the degree of freedom to act and the mental and business discipline that guides an organization.

The control system has been delineated as a separate activity because it involves more than the traditional approaches to financial control. Most business controls occur after the event has occurred. They seldom prevent an act from occurring. They report the history. Whether it involves reporting product costs or profits, the sources that produced those conditions are a *fait accompli*. Control in the sense here intended includes feedback loops as normally associated with process control systems—in any system there may be thousands of control loops. A properly instrumented chemical process makes adjustments to the system as it progresses to guarantee a product within certain specifications: Some raw materials may be outside the required specifications, but the end product can meet requirements if the system is designed to accommodate the variations by controlling other parameters. The process does not run to the end of the cycle without any correction.

The management control system must function in the same way. Managing the business system according to the principles of process measurement and control systems practice does not require expanding management information systems. Every organization will find that it is already encumbered by too much irrelevant data. Managing the business system requires determining what data is necessary and how those data can best be acquired and used to control the business process. Chemical processes are managed through a systems algorithm that guides the process within the limitations of the pre-established parameters. If the process parameters exceed the established limits, the process will automatically shut itself down. The management process also requires a system's algorithm, which means obtaining real-time information within the requirements of the process. Real time does not necessarily mean additional computerization; it does include providing a means for instant feedback and correction of the critical parameters. Managers must learn how to use feedback to correct the activities at the time the deviations occur in the management cycle and not wait until the end of the cycle. It is necessary to recognize that changing to a dynamic time-oriented organization is equivalent to managing the turnaround of a declining or bankrupt business. There are no easy answers to the complex questions.

The information in those many feedback loops must be constantly examined and used to modify the system. Figure 4.2 shows the two-way feedback loops that affect any business. If management adopts the type of thinking used in the chemical process analogy, whereby the process parameters are altered only when certain predetermined conditions are not met, it can make the necessary corrections at the time of their occurrence rather than wait for the results of some arbitrary reporting period. Continuous control of the management process requires only one basic consideration: understanding of the spread between the acceptable and unacceptable limits for meeting requirements. Encumbering a process with overly tight or narrow tolerances usually results in inordinate amounts of waste. A process can also be so broad and sloppy that nothing is ever produced to standards. If it is, it is late, costs too much, and does not do what was intended. Establishing the limits that provide the environment in which the use of the resources is transformed into acceptable products requires that executives and managers understand and use the information in the feedback loops to keep the system at optimal performance. That feedback process requires appropriate action.

Measurement is an integral part of feedback. It can be argued that the measurement of process parameters is considerably easier than the measurement of the parameters for managing the activities of people. Obviously, the accuracy of measuring input and output is considerably greater than the accuracy of measuring the response of people. That does not mean the latter should not be attempted. Determining what talent and level of talent are required for a specific project is not a mystery. Individuals either possess the necessary education, experience, skills, and attitude or do not. Past performance is a fact. Individuals who have consistently failed to meet commitments in the past, regardless of discipline or level, are a known quantity. Assigning a chemical engineer to design an electromechanical device may be a lack of good judgment. Project managers either do or do not have the ability to manage a project from concept to commercialization. A manager may be capable of managing a group of creative researchers, but that same person may completely fail in managing a group of product development engineers.

Lack of a business system perspective throughout the organization has not only caused businesses to lose competitive advantage but is responsible for the daily reports in the press about employee terminations. The recent disclosures by IBM [3], Xerox [4], General Motors [5], and other businesses regarding downsizing and terminations clearly show a *reaction* to current economic conditions. Although no one can be expected to predict the future, the executives in those organizations managed as though tomorrow would be the same as today. Inaction and reaction took precedence over a proactive approach. In the process, they neglected to manage the system.

Managing system cycle time, like managing a chemical reactor, requires a process. Process, in most organizations, is seldom considered beyond the manufacturing function. Yet it is the most underestimated and most undervalued management tool for optimizing the use of resources. Emphasis on process only can inhibit performance, but emphasis on *process and the outcome* enhances the output significantly. The outcome depends on how the process was described and how it is managed. Understanding the process and managing with the use of process documentation determine levels of productivity. Personal experience shows that manufacturing cycle time is often extended because of inadequate process documentation and follow-through. Omitting some inconsequential element in the process can escalate the costs of waste and scrap. Personal experience also shows that cycle time is extended because management fails to apply the fundamentals of process control to the management process. Managing the management process allows an organization to identify the system inconsistencies and discontinuities and take the necessary steps to correct the deviations.

Executives often ask: Why did the project related to product A meet all the requirements, and why was the project related to product B so unsuccessful? Why were the three elements of cycle time (total time, timing, and cycle duration) acceptable in one case and not in another? The answers to the questions are simple: Little if any attention to process and how it affects performance. Managers must focus greater attention on process, because that is where most of the problems begin. The primary purpose of understanding the process is to allow managers to interject the right amount of mental and operational discipline before the problems occur. Why did the Challenger disaster occur [6]? Why did the Hubble telescope malfunction [7]? The failures can be traced to the management process control system. That process includes communication among various levels of professionals and management. In the Challenger and Hubble examples the limits of operation were not established. It makes no difference whether that process relates to research, development, manufacturing, marketing, the other functions in the tripartite organization, or the business system. In this context, an appropriately designed management process establishes the safe limits of operation in the same way that the predetermined acceptable operating tolerances in a chemical process guarantee specific results. The limits on the operational parameters guarantee that the resulting product will meet not only the customer's requirements but also the business operational requirements. Managing, like the chemical reactor, requires a process. That does not imply that process is the only consideration. The chemical reactor requires materials that meet specifications; similarly, the management system requires resources that meet specifications.

Managers can help themselves meet the process requirements by resurrecting the old and forgotten principles of block diagramming, mapping, and flow charting. Every functional group must look at the process for achieving its results. As with the example of the chemical process, the intention is to continually modify or fine-tune that process to optimize the percentage of the input that eventually arrives at the customer's door. If a business is considered as a system similar to the chemical process, the actions and interactions must be optimized. In a multiple-reactor system, optimizing the output of one reactor will not achieve the optimal result. Likewise, in managing a business, optimizing one function will not optimize the output of the business.

Developing some form of process flow diagram for the critical business operations allows managers to combine and eliminate activities that add little or no value. Simplification is one of the key requirements in the design of products, processes, machinery, and services. Just as design engineers attempt to reduce the number of parts in an assembly or reduce the number of steps in a manufacturing process, executives and managers must apply the same philosophy and reduce the number of people in the cycle, the number of steps in accomplishing an objective, and the use of resources. Every unnecessary person, step, or resource adds cost, limits the opportunities for optimizing the system cycle time, and prevents allocation of critical resources to other more productive activities.

Rummler and Brache, in "Managing the White Space on the Organization Chart" [8], see a business organization as a system of horizontal processes rather than a collection of independently managed functions. They further suggest that the major variable in establishing a quality, cost, or cycle time advantage involves managing the cross-functional work processes. Managers must understand the role of process as contrasted to the more prevalent role of function. Rummler and Brache suggest that this occurs when:

- Key processes that cut across functions are identified
- The functional relationships and joint actions that are required to make the processes effective are determined
- The processes are measured and managed

Limits of the Business System

The question that must be answered is: What are the limits of the system? How the system is described determines the number of factors by which it will be influenced. As an example, with the discussions and urgings by the gurus for "going global," businesses must consider whether going global is required and, if so, determine the potential benefits. In most cases, businesses tend to describe their spheres of influence (the systems) and the other spheres by which they are influenced too narrowly. That approach can limit growth and also cause the demise of a business as competitors become dominant players. However, describing the spheres of influence too broadly, without an understanding of the fundamentals that guide the entrance into global markets, also can result in significant losses. Expanding into markets, whether national, international, or global, is a new ballgame and requires different talents.

The description of the system, that is, the limits of the ultimate boundaries, must be guided by the availability of the resources for exploiting opportunities within the system. Those resources include people, intellect, information, culture, customers, time, technology, finances, plant and equipment, and facilities. However, taking the systems approach requires a change in thinking processes.

R. L. Ackoff [9] describes a system as a set of two or more elements that satisfy the following three conditions:

1. The behavior of each element has an effect on the behavior of the whole.
2. The behavior of the elements and their effect on the whole are interdependent.
3. However subgroups of the elements are formed, each subgroup has an effect on the behavior of the whole and none has an independent effect on it. To put it another way, the elements of a system are so connected that independent subgroups of them cannot be formed.

If the foregoing description of a system is accepted, a logical conclusion would be that a system, although made up of individual parts, must be treated as a unit. Analysis must include the parts, the interrelations between parts, and the interrelations of the parts to the system. As Ackoff notes, every part of a system loses certain properties when separated, and every system includes properties that are not present in the individual parts. The concept should not be difficult to understand. We live in a world of systems beginning with our own bodies. Our five senses are part of that system, and loss of any one of them limits the activities and functions of the individual: the total human system. A similar statement applies to the world of technology. As an example, an automobile as a mechanism is a system. It also depends for its performance on a much larger system that includes oil drilling operations, refinery operations that provide the fuel, the fuel transport system, the network of service stations, the infrastructure of roads and bridges, a human driver, and the environment in which it functions. If the battery of an automobile is defective and prevents the engine from starting, the fact that all of the remaining parts may be in working order is of little consequence. Similarly, the vehicle might be prevented from operating because of environmental conditions.

Another example at a more complex level further illustrates the vast spectrum that systems can involve. The nuclear power industry certainly described its system in very narrow terms: building nuclear power plants. It disregarded environmental conditions, communicating intelligently with society about the potential safety hazards, transporting and storing nuclear waste, and resolving the issues related to the man-machine interface. It even disregarded or discounted technology limitations on its design process and its ability to control the results of the application of that technology.

There is no prescription for determining the limits of the business system. The system must include all the forces that in any way affect operations and the ultimate performance and sustainability of the organization. The system of the local machine shop will be quite different from that of a major equipment producer. The system of a conglomerate is quite different from the system of a major proprietary product supplier. Managers must carefully evaluate the forces that affect their performance and describe the system in which they function. It is an exercise that is absolutely essential for effective operation.

Systems thinking

A significant amount of the total time available to an organization is spent in two areas: (1) identifying problems and resolving them and (2) identifying new opportunities and providing the resources for their realization. Many books about problem solving have been written. The authors always assume that the problem has been identified. However, identifying the conditions or circumstances under which the problem may occur may be far more important. The problem and its sources must be clearly identified. Very little management time is spent on anticipating, not necessarily problems, but situations that may affect business performance. The word "problem" as used here means any deviation from the expected results whether related to production or any other activity in the organization. A parallel with the chemical process can once again be drawn. Chemical process control systems, as well as many others, are capable of anticipating process changes and making the necessary corrections so that the expected result can be attained. Managers must operate according to the same principles.

The differentiation between cause and effect often directs the problem solvers in the wrong direction. The causes are seldom clearly defined, and the solutions are more or less snap judgments rather than clearly thought out solutions requiring gathering of information for resolution of the problem. How many hours are wasted by jumping to conclusions that then require reevaluation after reevaluation until the problem is finally resolved? The same cause and effect considerations apply to identifying new opportunities. Problem definition and resolution and identifying new opportunities require analysis and synthesis. The analysis is only the first part. It involves confronting the facts rather than the opinions. Synthesis emphasizes understanding of the facts.

Balancing analysis and synthesis

Taking a systems approach requires a change in the mode of thinking. As noted, analysis receives primary attention from executives and managers. As a matter of fact, much of management has been dominated by analysis to such an extent that someone coined the phrase "paralysis by analysis." Much of the scientific and engineering community also focus excessive attention on analysis. Synthesis, on the other hand, is the orphan. Analysis and synthesis are different sides of the same coin, and their relationship cannot be changed. The approach of the past has been to take *things* apart, whether related to technology or business or people, analyze them, and then put them back together. Reassembly of the pieces takes place after some modifications have been made to the parts.

Sometimes that process may result in a new configuration of the thing. Use of the words "thing" and "things" includes physical embodiments of technology, concepts such as those related to marketing, actions by people, and any ideas that may be introduced to the organization in any situation. This analytical approach has been based on the fact that if all of the parts meet the requirements of some predetermined ideology or specification, the system will function as intended when all of the parts are brought together. Unfortunately, this theoretical approach is seldom found in practice, especially if the objective is optimization of the system. As an example, experienced technology specialists understand that an optimized technological system does not result from dividing a project into what are considered logical parts, assigning the parts to independent development groups, and then assembling the individual parts into a system without considering the interactions among the parts during the development process. That is true even when specifications for each part of the project are clearly described. The same provisions apply to projects that are organized by executives and managers related to the use of or optimization of the business resources.

Executives find themselves in similar dilemmas when they attempt to resolve organizational issues. As an example, reorganization appears to be a continuous process in most organizations. Executives continue to ask themselves: What happened? Why didn't the last reorganization yield the expected results? Why isn't the organization functioning as we thought it would? The answer is usually simple: No consideration was given to the system. A decision was made without understanding the implications of that decision and its impact on other parts of the system, on the whole organization, on the use or misuse of the ten business resources, or on the outside influences.

Thinking in terms of systems requires a reversal of perspective. In the past, attention has centered on analysis to the exclusion of synthesis. That does not suggest that the future will be dictated solely by synthesis. By itself, analysis has not proved to be a sustainable benefit. Changing to an approach totally based upon synthesis of questionable and unverified data will likewise not provide a sustainable advantage. Both always were and continue to be essential for effective and efficient management. Systems analysis generally includes:

1. Breaking the system down into its component parts
2. Gaining an understanding of each of the individual parts
3. Knowing how the different parts interact
4. Recognizing the contribution of each part to the system
5. Putting the system back together based upon what was learned

Those five basic steps apply to resolving all management issues whether related to problems or seeking new opportunities. They affect the local business that limits its products or services to the immediate community and the global megacorporations.

Steps 1 and 2 are usually achieved without any major difficulty, but steps 3 and 4 are seldom considered or receive only cursory attention. Discounting the importance of steps 3 and 4 when related to process or product results in dismantling of the system, redesigning and rebuilding parts or, more often, complete subassemblies, followed by assembly and retest of the total system. When related to management, continuous re-organization is the result: Each new reorganization is expected to achieve what the preceding ones failed to accomplish. Step 5 is really synthesis, but only if steps 1,2, 3, and 4 were performed diligently and in depth. "In depth" does not mean excessive amounts of study. It does mean concentrating on the major factors that affect the system's performance. Synthesis is no more than a fusion of all the known elements into a coherent whole. It requires bringing information together from many different sources and disciplines, evaluating that information in the context of the interactions, and putting that information together in such a way that the interdependencies of all the parts are satisfied. It means capitalizing on the total knowledge and expertise available in the system.

Analysis takes the whole and breaks it into the many individual parts. The synthesis approach considers the whole within the boundaries of the described system. An example may help clarify the difference between the analysis and synthesis approaches. For simplicity, assume that the research, development, manufacturing, and marketing functions are the four key functions in any business that are involved in bringing products to market. Label the group product initiatives (PI). Usually each of the four operational functions is further subdivided into many subfunctional groups. An analyzing mentality, in attempting to explain PI, would break PI down into its functions and subfunctions. Attention would be directed toward how each part of the PI system works. Then the need for management, professionals, and other participants would be delineated. From that analysis the purposes and objectives of each function would be described and result in a description of PI.

Approaching the same activity from a system's perspective would require identifying PI and all the factors relative to the expectations—a complete and concise description of the purposes, objectives, and strategies as noted in Fig. 4.2. That could include such considerations as the availability of the ten resources, the competitive issues in all phases of the business, pertinent industry standards, vendor and customer relations, developments in pertinent technologies, economic conditions, environmental considerations, community attitude, factors related to government regulations and its attitude toward the industry, and so on, with any specific issues peculiar to the business. All those factors become part of the system; all affect business performance.

Synthesis also looks for responses as to why things are as they are and uses the information to develop or reconsider the purposes, objectives, and strategies of the organization. It is concerned with interactions rather than actions alone. The roles of research, development, manufacturing, and marketing in the PI example can now be described quite differently. It is conceivable that in certain organizations the department now called research would be eliminated. That does not suggest that expenditures for research would be eliminated, but research as a department or a specific function may have outlived its usefulness. The research activities may be integrated into the PI structure and lose their identity.

Over the past two or three decades, the technology functions, primarily those involving scientists and engineers, have been transformed into a multiplicity of independent kingdoms. Contrast the current situation with the following example. My first position, after graduating with a degree in electrical engineering, involved assignment to what was known as the technology department. The department included about 200 scientists, engineers, technicians, and support people. There were no distinctions between research, development, design, manufacturing engineering, tooling, and so on. No individual kingdoms. Research focused on the research necessary for the expansion of the company's product line and businesses. It was driven by the needs of the business. Engineers who worked in design knew that the products they designed would have to be manufactured at competitive prices. Designers were provided with target cost requirements. No one had to ask if the designs could be manufactured cost-effectively. No one raised the questions about the impossible tooling requirements. There was no question that tooling would be required and that it must be provided at costs that were consistent with the production quantities. The three functions—research, development, and manufacturing—were one. Marketing, the fourth function of the PI example, was integrated into the system by appointing a vice president for marketing and technology. No longer could marketing or research and development play the games that are usual when one or the other dominates the scene. That does not suggest that businesses should reorganize their technology and marketing functions along such lines, but it does demonstrate the concept of an alternate solution for a specific situation.

In the 1990s, managers continue to be concerned that product designs cannot be manufactured competitively because of an excessive number of parts, difficult assembly operations, countless engineering changes, and so on. What changed? There is no doubt that technologies may be more complex and that projects need talent from a variety of disciplines. But the tools to accomplish these more complex technologies also are available. In the past two or more decades the technology functions have been segmented to unworkable minifunctions and have become so specialized that each individual focuses only on the requirements of her or his own specialty. If a problem arises, many different disciplines must become involved because of the microspecialization. It is difficult to find an individual who understands the workings of the system. In the example cited, there was a department called technology. It included chemical, materials, mechanical, flow systems, and electronic research. This department was also responsible for product design and testing, tooling, manufacturing engineering and all the related functions such as pilot plant and prototype construction. It was impossible to work in such an environment and not learn something about the other disciplines. The managers made it a point to provide a variety of assignments so that although one's specialty was of primary importance, it did not overshadow the need for some breadth of technological understanding and expertise. The department was organized according to projects. No staff specialists were assigned as consultants. No individual was a *responsible-for* specialist, who may be the individual responsible for technology transfer, patent liaison, various coordinating functions, or international technical liaison. Individuals were expected to make a contribution through some specific activity or series of activities based on well-developed objectives. This technology department considered all of the elements of the system from concept to commercialization. Communication problems were essentially nonexistent.

Analysis and synthesis are two fundamental precepts in problem resolution. Uniting them in the business decision-making process allows managers to ferret out the bottlenecks in an organization and then provide a rational base for developing conclusions based on the needs of the system. It is clearly not a matter of either analysis or synthesis; it is a matter of both. The two are complementary activities. As Ackoff suggests, analysis yields knowledge and synthesis yields understanding, which should provide for more effective and efficient use of resources. Analysis looks into things; synthesis looks out of things.

Analytical thinking is concerned with the interactions of the parts. Synthesis as a necessary ingredient in systems thinking is concerned not only with the interaction of the parts but also with the interaction of the parts and the system and the other external elements of the environment in which the system operates. The limits and the boundaries of the many environments must be specified clearly.

Managing the System

How can the system be so managed as to optimize the three elements of cycle time? Business executives and their managers must become systems-oriented. They have thought and continue to think about their management functions as separate and independent activities rather than as a unified system. Managing the pieces takes precedence over managing the system. Systems thinking is not a new concept. Jay W. Forrester, in the classic article "Industrial Dynamics: A Major Breakthrough for Decision Makers" [10], presented some fundamental principles:

> Management is on the verge of a major breakthrough in understanding how industrial company success depends upon the interaction between the flows of information, materials, money, manpower, and capital equipment. The way these five flow systems interlock to amplify one another and to cause change and fluctuation will form a basis for anticipating the effects of decisions, policies, organizational forms, and investment choices.

Forrester also attempted to redirect attention from emphasizing techniques and prescriptions toward developing a professional approach to management.

> Business leaders, like leaders in other areas, are influenced by an *image of the future*. Their ideas about where they are going may be clear or vague, but in any event they have a subtle and far-reaching impact on administrative thinking and decisions. A look at some promising new concepts of management should, I believe, convince even the skeptical executive that his job is developing into much more than an art, that conceptual skill will play an increasingly vital role in company success, and that management is fast becoming second to none as an exciting, dynamic, and intellectually demanding profession.

Forrester's comments were published in July 1958, not 1991. Managers continue to manage independent functions rather than the system. Instead of expanding the system to include the external forces acting on it, they have in reality looked inward and further segmented the internal functional groups, through overspecialization, with the hope of achieving greater efficiency.

An executive at the Honda Marysville, Ohio plant [11] is reported to have said: "When it comes to management, Americans do not practice what they preach." He went on to say that the United States has progressively dismantled and discarded the principles of management that Americans taught to the Japanese, and in the process they lost their competitive advantage. The concepts of just in time (JIT), concurrent or simultaneous engineering, merging the activities of product development and manufacturing or other business functions, and continuous improvement are not new. They were introduced and practiced in the past by U.S. industry in one form or another.

No additional research is required to verify the Honda executive's statement. United States managers are rediscovering what their predecessors began. A very large percentage have ignored the necessary business discipline, the need for operating under some unified set of principles, and need to take advantage of the lessons learned from the past. It is only necessary to look at the best seller list of management books to understand that executives and managers focus attention on the quick fix and the "ten easy steps to improved business performance" panaceas. They discount the benefits of following the fundamental principles and practices of management.

Hayes and Abernathy, in their 1980 article in the *Harvard Business Review* [12], noted that "modern management principles may cause rather than cure sluggish economic performance." They proposed that American managers have not directed their attention toward the role of technology in the marketplace: competing by offering superior products. They state:

> Guided by what they took to be the newest and best principles of management, American Managers have increasingly directed their attention elsewhere. These new principles, despite their sophistication and widespread usefulness, encourage a preference for (1) analytic detachment rather than the insight that comes from "hands on" experience and (2) short-term cost reduction rather than long-term development of technological competitiveness.

There is little argument that managers increasingly rely on principles that promote analytical detachment and methodological elegance over insight. We are reminded of that on a daily basis in every business publication. Experience is discounted when considering the subtleties and complexities that are involved in the management process. Hayes and Abernathy claim that this approach of analytical methods only, working from paper rather than hands-on involvement, overemphasizes the short-term financial returns and underemphasizes the long-term future of the organization. Although it is common practice to criticize industry for its short-term approach, academics like Abernathy and Hayes should also recognize that long-term success can be accomplished only by meeting the short-term goals. A balance between the resources allocated to short-term and long-term objectives is essential. The balance would be quite different for a company like Hershey Foods and Hewlett-Packard.

Hayes and Abernathy also note that, with this short-term view, pseudo-professionalism has been idealized and has taken on the "quality of a corporate religion." Its precepts are simple: (1) Neither industry experience nor hands-on technological expertise counts much. (2) As issues rise to the top of the managerial hierarchy, they are distilled into easily quantifiable terms. An adequate response to the first issue would evolve into a lengthy study. In recent decades, however, U.S. industry has become credentialized. Academic standing, position, and so on take precedence over responding to a simple question: What were the individual's personal contributions to the business? There is no research that correlates academic achievement with business performance. Attitude, personal drive, creativity, and meeting commitments cannot be predicted or measured by scholastic aptitude. Multiple choice and true or false tests are not indicators of future performance.

In regard to point 2 made by Hayes and Abernathy, there is no doubt that simplicity is an absolute essential, since the process of simplification brings understanding. Distillation of data into information and knowledge is the product of management. It is also the only way to educate the illiterate, whether that illiteracy involves technology, marketing, or business. However, reducing to the fundamentals, to the basics, and describing a complex concept or situation in simple terms requires hard work. It also requires a great deal of proactive thinking. Simple in that sense does not mean simplistic; the two words are totally different. "Simple" implies reducing complexity to its essential elements and communicating the results in intelligible language—eliminating the professional jargon. "Simplistic," on the other hand, indicates a lack of understanding of the various interrelations, inadequate effort directed toward reaching a conclusion, and disregard of the fundamental principles associated with simplifying: thinking thoughts, making associations, analyzing, synthesizing, and repeating the process as often as necessary.

Ackoff, in the preface to *Creating the Corporate Future* [13], says:

> A good deal of the corporate planning I have observed is like a ritual rain dance; it has no effect on the weather that follows, but those who engage in it think it does. Moreover, it seems to me that much of the advice and instruction related to corporate planning is directed at improving the dancing, not the weather.

Ackoff's comments can be paraphrased in relation to the general issues associated with managing the business system:

A good deal of corporate managing is like a ritual rain dance; it has no effect on the business performance that follows, but those who engage in it think it does. Much of the advice and instruction related to managing is directed at improving the dancing, not the business performance.

During the last two to three decades business lost sight of the basics and the fundamentals of managing. People needs, or, more appropriately, wants, were overemphasized. The principles that made U.S. industry the envy of every nation were disregarded. They were assaulted by the academics and the management researchers who never managed. In their impatience to criticize the contributions of such people as Frederick W. Taylor, Henri Fayol, Frank and Lillian Gilbreth, and Henry Gantt, without first gaining some understanding, they ignored the differences between principles and the application of the principles. There is no doubt that Taylor's principles were misapplied many times, but that does not negate their essential usefulness. An educated emphasis on efficiency could change the future prospects of the American economy. Those fundamental principles of management expressed by Fayol, namely, planning, organizing, staffing, directing, and controlling, although insufficient today, are nevertheless absolutely essential. Additions to the five principles would include measuring and/or monitoring, integrating, coaching and educating, communicating, personalizing, pushing and pulling, analyzing and synthesizing, reviewing, meeting commitments, and two of the most important, doing and participating [14].

In this same period, in which executives and managers ignored the fundamental principles, the principles were blindly replaced by an employee-dominated needs mentality. Employee needs must be balanced with the other needs of the business. That idealized view brought into being the concepts related to self-managed groups, inverted hierarchies, total freedom, team building, empowerment, and so on. All those approaches may provide some benefits to the organization, but success depends on how they are implemented. Recognizing that Utopia does not exist, someone must be responsible and accountable for performance. People are not ideal creatures. They come with all of their talents and capabilities as well as with all their foibles and limitations. People cannot be empowered without education. In the past, organizations only asked for loyalty, faith in the decision makers, obedience, and a modest level of competence. Accepting personal risk and sticking your neck out received no emphasis from human resource management. At the same time, no research demonstrated that providing the people-oriented programs benefited the individual or the organization. Corporate executives allowed their managers to focus exclusively on people needs, including their own, rather than on business needs.

An outcome of this mental attitude of executives, managers, and all the participants was the total dependence of the individual on the organization. One is reminded of a reenactment of the birth of the industrial towns in which companies provided the housing, the company stores, the schools, and so on. How professional employees could place their entire future in the hands of some corporate structure appears to be unbelievable and unjustifiable. How an intelligent human being can delegate that responsibility to some entity such as a corporation is incomprehensible. Executives, through their actions and corporate policies, gave birth to a community of managers and professionals who foster an idealized and fictional view of just what contribution an organization has a right to expect in return for compensation of its employees.

If organizations intend to manage their business system cycle time, the current employee-need mentality must be replaced with one that focuses attention on performance and recognition and compensation for superior performance. The levels of expectation must be raised not only for the benefit of the organization but equally for the benefit of the individuals. How can an organization tolerate professional obsolescence in any discipline? How can an individual tolerate personal obsolescence? Those issues must be addressed, and change in the current organizational cultures will not occur through edict. It will require the best of available leadership that practices what it preaches.

Ulrich, in "Competing from the Inside out" [15], suggests that what happens inside an organization with people and processes will affect what happens outside with customers and other stakeholders. Executives should not need any quantitative research results to demonstrate the validity of this truism. It is a given, but it is generally ignored. Technological, financial, and marketing competence are essential for success but are insufficient by themselves. They are moving business targets, and their benefits for competitive advantage are short-lived. The competition quickly plays the game of catch-up. System cycle time will not become a reality and a business advantage until, as Ulrich describes, leadership at all levels of the organization is in place. That leadership requires shared responsibilities for business success, employees with a sense of ownership, willingness to take risks, breaking new ground, and the stick-to-itiveness for accomplishing what everyone else said could not be done. Ulrich uses the metaphor of the pioneer to describe the participants: people who are highly individualistic but dependent upon the team for success, who act without full knowledge of the challenges that lie ahead, and who rely on their core values in difficult times.

Strategic and Operational Issues

Focusing attention on system cycle time determines the subsequent cycle times associated with products and projects. It establishes the foundations on which the organization is built. If communicated throughout the organization it results in a culture that balances the needs of all the stakeholders, both within and outside the organization. The elements of the infrastructure establish the direction in advance of the need. They allow the organization to be sensitive to the environment and to participate in creating change rather than be dominated by the actions of others.

Managing system cycle time requires consideration of the strategic and operational issues related to developing what has been described as the infrastructure of the organization. The infrastructure issues although qualitative—since no fixed algorithm that would measure their impact with any degree of precision can be generated—nevertheless make a significant difference in performance. The elements of the infrastructure must impact, in some positive way, the manner in which all the participants work. Some may wish to describe this infrastructure as a cultural issue. But it goes far beyond the simplistic descriptions of culture that are limited to the myths, shared values, beliefs, legends, heroes, rituals, and other artifacts. To be of value, culture must somehow change behavior. Talking about the heroes of the past or reviving the old war stories has its place, but if the elements of that infrastructure (culture if you insist) are to be effective, behavior patterns must be changed.

Strategic issues

The following comments on strategy are made for reference only. This is not a detailed discussion of strategy and its implications on business performance [16]. The available literature on strategy, strategic planning, and strategic management presents a never-ending stream of comments from the academic as well as the business community. Strategy, not strategic planning, is important, but it depends on the description. In simple terms, strategy tells us how we are going to get from where we are to where we want to be. However, arriving at the resultant strategy and the subsequent activation of that strategy in the system is a complex process. Strategy, very simply, provides the direction. It tells an organization, how, in very general terms, it will go about accomplishing its purposes and objectives. It is not a detailed plan or a book of competitive statistical data; it is a relatively simple statement. As an example, one of 3M's strategies for meeting its growth and profit targets considers the introduction of new products—25 percent of sales in any one year must come from products that were not available within the preceding 5 years. That is a very simple and definitive statement—a strategy that describes how the growth targets will be met. How each operating unit within the organization accomplishes its objectives in contributing to corporate purposes and objectives will require additional individualized strategies based on the description of the specific business unit.

What strategies should be used to optimize the segment of cycle time from point X to point Y in Fig. 4.1? The word "optimized" is used instead of "reduced," "shortened," or "minimized" as is generally suggested. It is a matter of the right cycle time, not the fast or the slow, but the cycle time required by the system. At times it may be necessary to reduce the element of cycle time and at other times it may be appropriate to extend it. As an example, very deliberate and sound reasoning can lead to a decision to delay any major investment for some period of time. But if that decision is made, it should be communicated and the effort should be redirected toward some other activity. Perhaps some continued scouting may be required, but if necessary there should be some reasonable expectations from that effort. It should not be allowed to linger in limbo, where valuable resources continue to be expended without any measurable benefit. It may be appropriate to reconsider the investment at some future time; if so, that should be specifically stated. Likewise, it may be necessary to change priorities because of the dynamic nature of the business and apply all the resources to some single and very specific end. These decisions must be conscious and must be made only after consideration has been given to the net impact on the total system.

Before a strategy for dealing with the system cycle time can be developed, some knowledge of the lost time and its impact on business performance is essential. Origins of lost time were discussed in Chap. 3. Without that knowledge, there can be no strategy. A baseline must be established. That is a difficult task, not from the standpoint of complexity, but because a lack of objectivity and integrity distorts the reality. It requires facing up to some past poor judgments and lack of attention to detail and admitting to some avoidable failures while capitalizing on the lessons to be learned from those experiences. This is not a witch hunt; it is not a means for laying blame. It begins with the CEO and exempts no one in the organization. The discussions must be objective, and personalities cannot enter into them. It is necessary to keep in mind that every individual in the organization is part of the system, and that, through silence, naivete, apathy, or disinterest contributes to system cycle time.

The strategy to optimize the system cycle time can be stated very simply. It refers back to the description of strategy: How does an organization go from where it is in lost time to where it wants to be? Rephrasing, how does an organization optimize the use of its resources to optimize the system cycle time? The components of cycle time, which include the three types of cycle time, the five classes of time, and the three elements of cycle time as noted in Fig. 2.8, establish limits on business performance and pervade every business-related activity. Three self-evident truths must be accepted if a workable system cycle time strategy is to improve business performance:

- Lost time can never be regained.
- Missed milestones add cost throughout the total system.
- The cycle time must be appropriate to the demands of the specific situation.

The strategy to optimize system cycle time requires only a simple statement: find the sources of lost time and eliminate them.

Operational issues

How can that strategy be implemented? A simple approach would include such activities as creating awareness, establishing companywide educational programs, and following up with very specific objectives or targets. But is that sufficient? If it is, what does it really accomplish, and what is the net result? There is no doubt that awareness and education are essential activities in the process of creating change but are insufficient by themselves. The right people must be in place in order to create the change, take advantage of the benefits of that change, and then capitalize on the synergies that can be exploited. The time-to-market and the speed proponents and enthusiasts fail to recognize that to make a major impact on cycle time requires a change to a *time- and objective-oriented organization*. It involves more than saying: We have to do it faster. Certain non-value-adding activities must be eliminated. Persuading people to speed up their work by doing a non-value-adding activity faster provides no benefit. The system that must be in place allows the optimization of the ten business resources, and time is one of them. It has a value.

If an organization attempts to exploit system cycle time management as a means of gaining competitive advantage, it must change the way it thinks. It must think cycle time in terms of:

- The dynamic rather than the status quo
- Integrated systems
- Holistic management

- Related and useful information that is verified as to its integrity
- A continuous flow of ideas that support its purposes and objectives
- Well-developed business concepts that can be tested and evaluated
- Innovation in all business operations within the ability to utilize them
- Timely decisions in all matters
- Optimal use of all resources

That may sound like an impossible task, but that's what managing is all about.

Changing the thinking to a time- and objective-based organization is no simple task. Any change tends to disrupt the work flow and the peace and quiet environment that many organizations have fostered in the past several decades. System cycle time management is important to the continued growth of any organization. By itself, growth is never the dominant objective of a business, but neither is maintaining the status quo. Exploiting cycle time through effective management practices must go beyond the simplistic approach of applying new tools. It involves reducing the allotted time for accomplishment, optimizing the cycle time, and focusing attention on the proper timing. It requires a *new think* by all of the participants.

There is no operational prescription for developing a dynamic time- and objective-oriented organization. There is a simple methodology, though, that can be applied in order to begin the process, and it is no different than any problem-solving process. It involves reflecting to some extent on past performance, but it places greater emphasis on looking at the present. That process involves gathering the necessary inputs, processing the inputs, and reaching some output conclusions. More specifically:

- Identifying the sources that consume time without adding value
- Describing the sources of lost time in sufficient detail
- Analyzing the information as to origin, type, cross functional impact, and so on
- Formulating a statement of the causes
- Developing an action plan

Summary

1. System cycle time management requires consideration of the system, management, and technology. Those three elements are essential for developing a dynamic time- and objective-based organization.

Forrester described the need to manage the interaction of the flows of information, materials, money, manpower, and capital equipment, which have been expanded and are part of the ten essential business resources mentioned. He also suggested the need to anticipate effects of decisions, policies, organizational forms, and investment choices. Ackoff's comment about planning the rain dance was paraphrased in terms of managing to caution against improving the dancing to the detriment of business performance.

2. Process influences business performance and cannot be ignored. It sets the limits within which an organization functions and yet stays in control of its business. If businesses had focused more attention on process in the past, perhaps *The Wall Street Journal* would not be providing us with a continual recitation of negative results. Monitoring the process allows for timely feedback to be acted upon in advance of the problem.

3. Ackoff sets down the conditions for describing a system. The system must yield something greater than the sum of its parts. Looking at the sum is insufficient. An organization that begins thinking in terms of developing a dynamic time- and objective-based organization must recognize the codependence of analysis and synthesis in managing performance: a truth self-evident but seldom observed. The process involves changing the perspective from which any problem is viewed as well as viewing the problem through different eyes.

4. System cycle time must be described to fit the organizational needs, but in sufficient breadth to include all the factors that impact its performance. The breadth will describe how the organization thinks about itself and the influences under which it operates. The elements of the infrastructure control the system cycle time and determine the probability of success in becoming a time- and objective-based organization.

System cycle time is both a strategic and an operational issue. Without a strategy for implementing system cycle time management, the negative impact of lost or wasted time, improper timing, and extended cycle times will continue to prevail. Managing system cycle time, which includes managing the factors in the business infrastructure, allows reaping additional potential benefits from improving product and project cycle time.

References

1. Y. Kathawala, D. Elmuti, and L. Toepp, "An Overview of the Baldridge Award: America's Tool for Global Competitiveness," *Industrial Management,* March/April 1991.
2. J. T. Vessey, "The New Competitors: They Think in Terms of 'Speed-to-Market'," *The Academy of Management Executive,* vol V, no. 2, 1991, pp. 23–33.

3. P. B. Carroll, "IBM is Likely to Trim 20,000 Positions," *The Wall Street Journal*, Nov. 20, 1991, p. A3.
4. K. Pritchett, "Xerox to Cut 2,500 of Staff, Take a Charge," *The Wall Street Journal*, Dec. 12, 1991, p. A3.
5. J. B. White, "GM's Problems Have Overtaken Stempel's Go-Slow Approach," *The Wall Street Journal*, Dec. 16, 1991, p. B1.
6. K. A. Kovach and B. Render, "NASA Managers and Challenger: A Profile and Possible Explanation," *Eng. Manage. Rev.*, March 1988, pp.2–6; reprinted with permission from *Personnel*, April 1987, pp. 40–44.
7. *Star Tribune of Minneapolis*, "One Step at a Time into Space," Dec. 16, 1990, p. 24A.
8. G. A. Rummler and A. P. Brache, "Managing the White Spaces on the Organization Chart," *Supervision*, May 1991, pp. 6–12.
9. R. L. Ackoff, *Creating the Corporate Future*, Wiley, New York, 1981, p. ix.
10. J. W. Forrester, "Industrial Dynamics: A Major Breakthrough for Decision Makers," *Harvard Bus. Rev.*, July-August 1958, pp. 37–65.
11. E. R. Archer, "Toward a Revival of the Principles of Management," *Industrial Management*, January-February 1990, pp. 19–21.
12. R. H. Hayes and W. J. Abernathy, "Managing Our Way to Economic Decline," *Harvard Bus. Rev.*, July-August 1980, pp. 67–77.
13. R. L. Ackoff, *Creating the Corporate Future*, Wiley, New York, 1981, pp. 13–24.
14. G. H. Gaynor, *Achieving the Competitive Edge Through Integrated Technology Management*, McGraw-Hill, New York, 1991, pp. 209–226.
15. D. Ulrich, "Competing from the Inside Out," *Executive Excellence*, June 1991.
16. For a detailed discussion of strategy see R. Vancil, "Strategy Formulation in Complex Organizations," *Sloan Manage. Rev.*, Winter 1976; J. B. Quinn, *Strategies for Change—Logical Incrementalism*, Richard D. Irwin, Homewood, IL, 1980; H. Mintzberg, *Mintzberg on Management*, Free Press, New York, 1989; M. E. Porter, *Competitive Strategy—Techniques for Analyzing Industries and Competition*, Free Press, New York, 1980.

Product Cycle Time

Product cycle time fits somewhere on the continuum from point X to point Y in Fig. 4.1. How product cycle time is described determines where points A and B are. For ease of presentation, the word "products" includes products and services and the processes by which the products and services are delivered. Product and process cannot be separated. The attitude that process becomes important only as a product reaches maturity is not now and never was valid. At the very least, the attitude resulted in lost profits over many years because of excess costs. Just because an organization meets its profit targets does not mean that the results were achieved effectively and efficiently. Product and process research and development begin at the same time, that is, if an organization is attempting to optimize business cycle time.

Optimizing product cycle time requires exploiting the ten resources available to the organization. Exploiting those resources requires redescribing and reconsidering some related management thinking and expanding some current and narrowly defined points of view. The following issues are considered and described in terms that meet the requirements in the current competitive climate:

- Scope of product cycle time
- From concept to displacement or replacement
- Integration of technologies, products, and markets
- New product description in relation to cycle time
- Product life cycle
- Information transfer from research to marketing
- Interaction of research, development, manufacturing, and marketing
- New product process assessment
- Ancillary considerations

Scope of Product Cycle Time

Businesses describe product cycle time in different ways, some of which are more inclusive than others. Usually, they limit the description as the period of time either from research to manufacturing or from development to manufacturing. Every organization must establish its own limits, but those limits should include the time from product concept to product displacement or replacement. Use of shorter periods of time causes an organization to lose sight of the decreasing value of current products. The benefits of broadening the description to include concept to displacement will become self-evident.

The literature on the various considerations for shortening cycle time often provides examples of how some organization reduced cycle time. The examples usually focus on some limited or very narrow operation within the total business system. Articles usually relate the following types of accomplishments:

- Design analysis was reduced from 2 weeks to 38 min (the goal is 4 min) [1].

- The number of engineering changes per drawing dropped from a high of 15 to 20 to a low of 1 [1].

- A manufacture-to-order business that once required a manufacturing lead time of 100 days managed to shrink it to just 3 days [2].

- Boeing Commercial Airplane Group [3] is using concurrent engineering to develop the giant 777 transport and expects to release design drawings 1-1/2 years earlier in the process than was true of the 767 [3].

- John Deere & Co. used concurrent engineering to cut 30 percent off the cost of developing new construction equipment and 60 percent off development time [3].

- AT&T adopted concurrent engineering and halved the time to make a 5ESS electronic switching system [3].

- Hewlett-Packard claims that it required only one-third of the normally required time from idea to finished product to develop the HP54600 oscilloscope [3].

Such statements must be questioned to determine the real results. In the examples cited, reducing a design analysis from 2 weeks to 38 min may or may not affect total performance. It must be looked at in the context of the total system. What was the cost of that 38 min? How much capital was required to accomplish it? What did the 38-min analysis include? Was the time saved used productively? What were the measurable benefits? In the second example, what type of management system allowed 15 to 20 engineering changes per drawing? The problem goes beyond reducing the number of engineering changes. Some generic operational and management changes are required in the system. Many of the engineering changes did not originate in engineering. Their sources must be determined, and actions to eliminate the lost time must be taken.

Consider a verbally reported situation in which an organization began the development of a lower-cost replacement product that included the design of the product, a new manufacturing process, new packaging, a search for new vendors, field testing, and so on. In the past, according to the executive, that type of effort would have taken a minimum of 15 to 18 months, but by using simultaneous engineering practices, the time was reduced to 75 days. The contribution to business performance and the validity of that and the preceding examples must be determined. What factors allowed reduction in time to market from 15 to 18 months to 75 days? How that product could have been designed, built, and field tested, including the redesign of the manufacturing process, in 75 days is difficult to comprehend. Similar questions must be asked of the other examples. Managers must make sure they are comparing apples and apples. Reality cannot be couched in internal public relations releases. All those good programs are useful only if they make a difference in the bottom line, and that involves optimizing the results of the ten business resources. That is the dilemma in reporting the experiences of organizations that attempt to show their progress in adopting fast cycle time. It is necessary to describe clearly what is meant by a new product and to establish the reference point against which advantage or progress is measured.

There is no doubt that some benefit was received from the actions cited in those examples, but probably at a minimal level in relation to the magnitude of the problem. Quoting percentages does not tell the complete story. The question that must always be answered concerns the adequacy of the improvement. Was it sufficient in the particular circumstances? What was the starting point? Consider the situation in which an operations group was boasting how manufacturing cost of a product was reduced from approximately $15 per unit to $4. The senior executives even awarded prizes to the manufacturing unit. By doing so, they sent the wrong message to the organization. The situation should never have occurred; there was really nothing to boast about. A series of bad technical and marketing decisions were made in the initial stages of the development. The manufacturing technology received only cursory consideration, and the scope of the market was completely misread. As a result, the initial manufacturing cost was 4 times the competitor's selling price. After introducing new manufacturing processes, which required extensive investments over several years, costs were finally reduced to competitive levels. But there was no real competitive advantage. The company used the same manufacturing technology and processes as its competitors.

There were no valid reasons for the bad decisions associated with the program. The information was available. The technology and marketing approaches had been critiqued by various specialists, and the inconsistencies had been called to management attention. The managers chose not to listen. Their manufacturing management knew nothing about manufacturing technology. Well-developed concepts of continuous processing that have been proved in practice were never explored. Manufacturing management became technologically obsolescent. All the related functions ignored the readily available information that could have made a significant difference in the outcome. Not only was the time for accomplishment multiplied many times, but the timing of the product into the market was delayed and the cycle time to market was extended. Percentages can be misleading; they can put managers into a euphoric state that ignores reality. Yes, reductions were made in manufacturing cost, but those costs should have been considered in the planning and product design process.

From Concept to Displacement or Replacement

To take advantage of all the possibilities in optimizing cycle time and exploiting it for improved business performance, it is essential that cycle time include more than the narrow limits normally attributed to it. The concept to replacement or displacement approach applies not only to the product itself but also to the activities and the work methods used in achieving certain specified results. A distinction should be made between replacement and displacement. "Replacement," as used in this context, means substitution by another product that has similar or better characteristics but provides essentially the same functions and results. "Displacement" means removal of a current product from the market because the product no longer meets the market need: It is obsolete and no longer serves a useful function.

Why describe the product cycle time from concept to replacement or displacement? Every product or service will pass through the usual stages from genesis to demise. In the early stages, as in nurturing a child, management places a great deal of attention on nurturing a new product to achieve the expected results. At some time, the attention given by management is inversely proportional to the profit contribution. That in turn leads to a feeling that the product will continue to produce without any additional investment of time or money, an attitude that usually results in a continual decline in sales and profits.

Someone eventually asks: What happened? The answer or answers are quite simple: Management fell asleep or at least took a long nap, or the product and all the activities associated with it were treated as what some strategic management gurus referred to as the *cash cow*—except that the cow no longer produced a sufficient amount of milk to justify its intake of food and upkeep.

Every business, whether a blue chip company or a more modestly managed organization, has its share of problems in regard to replacement or displacement products that usually result in very costly adventures. Managers do not think in terms of obsolescence. They continue to project the upward curve in sales and profits. Contrast that to the approach of Sony to introducing the Sony Walkman [4]. The first Sony Walkman stereo cassette player was introduced in 1979. It was followed by a series of modified cassette players and culminated in the introduction in 1985 of two versions of the Professional Walkman and a double cassette deck. Sony introduced the Solar-Powered Sports Walkman in 1986, the Fashion Walkman Radio Cassette Player and the Fashion Walkman in 1987, and the 10th Anniversary Super Walkman in 1989. It introduced the Recording Walkman, the Sports Walkman, My First Sony, and the Outback Walkman in 1990. Over a period of 12 years, many different models were introduced. No technological breakthroughs, no marketing ploys, just a continuous introduction of different versions to meet different needs—slight improvements in technology and addition of functions focused on specific markets.

Unlike Henry Ford, who was to have said that a buyer could purchase a Ford of any color as long as it was black, Sony provides an array of styles, colors, features, and so on to meet individual preferences. When will the Sony Walkman become obsolete? Probably not for a long time if Sony continues its past strategy of continually improving its products, adding specialized features, and obsoleting older models. The U.S. auto industry at one time, which seems decades ago, followed a similar approach. People looked forward to seeing the new models with the changes in design and function. Adding a fancy cup holder in today's new automobile entries does not seem very exciting.

An argument could be made that the Sony example applies to life cycle rather than cycle time. It applies to both. It applies to product life cycle insasmuch as the business strategy focuses attention on marketing a continuous stream of products based upon similar technologies and serving the same markets, although different niches. It is an example of how, through adding new features and benefits, older versions are obsoleted and replaced with units in which minor improvements provide new desirable features for a specific segment of the market. It applies to exploiting cycle time in view of the fact that such an infusion of products into the marketplace can be accomplished only by considering cycle time as the time from concept to displacement or replacement.

Integration of Technologies, Products, and Markets

Product cycle time must be integrated with technologies and markets. That may appear to be obvious, but most organizations continue to treat technology and market independently rather than as a unit. Products rely on technologies; products depend on markets. New technologies without product or market applications provide no benefit. New products without new technologies provide a minimal benefit. Products without markets, regardless of their internally perceived benefits, consume resources without benefit.

Figure 5.1 shows the three-dimensional relations of technologies, products, and markets. It is one way to show the relations and some of the complexities involved in attempting to introduce new products with new technologies, new products into new markets, and new products with new technologies into new markets. This chapter emphasizes product cycle time, whereas the holistic and systems approach requires that all activities related to the commercialization process be included. Product cycle time involves more than just the time to develop the product. The product must be manufactured at some predetermined

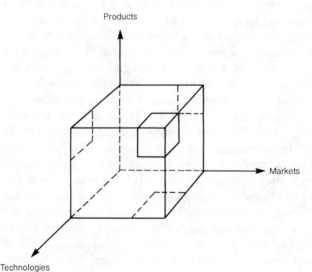

Figure 5.1 Relations of technologies, products, and markets. Each axis shows a continuum of current to new and increasing complexity from the point of origin.

cost, and all the means of production must be included. The product must eventually be sold, so all of the functions of the distribution entity must be included.

Figure 5.2 reduces the three-dimensional matrix to three two-dimensional matrices. It shows the relations of technologies and products (Fig. 5.2a), products and markets (Fig. 5.2b), and markets and technologies (Fig. 5.2c).

The three matrices of Fig. 5.2 have a common format. The horizontal axis and the vertical axis represent a continuum from current to new. Although the axes actually represent a continuum, they are divided into quadrants for a more simplified explanation. There is seldom a specific dividing line between current and new. As an example, in the case of technologies, it is doubtful that each and every technology would be new. Figure 5.2a relates technologies to products. The four quadrants of the matrix are designated 1, 2, 3, and 4. The move from a current product to a new product may be relatively simple. Likewise, the move to some new technology from quadrant 2 to 3 also may be accomplished with minimal delays. To go to quadrant 4, which would require new products and new technologies, is exceedingly more complex.

Although technologies are introduced to the marketplace through new products, an argument could be made that the relation between technology and markets shown in Fig. 5.2c is redundant. It is generally disregarded without realizing that adequate technologies, and not necessarily the latest or the most sophisticated technologies, must be implemented. Technologies such as those shown in Fig. 5.2a are generally governed by technical choice: They must also be governed by the specific markets to be served. New technologies can be introduced to existing markets with minimal risk. Current technologies also can be introduced to new markets with a minimal risk. If an organization chooses to enter quadrant 4, new technologies to new markets, the risk increases dramatically.

The matrix of Fig. 5.2b, which relates products to markets, requires consideration of the risk factor. Entering a new market with a current product, like introducing a new product to a current market, entails

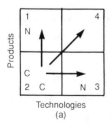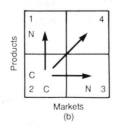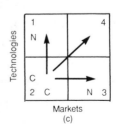

Figure 5.2 Technologies, products, and markets in two-dimensional matrices. C refers to current status and N to new status.

minimal risk. Most organizations should be able to move from quadrant 2 to quadrant 3 and from 2 to 1 with certainty. It requires nothing more than doing the homework and preventing egos from ruling the decision processes. A move into quadrant 4 with a new product to a new market presents an organization with considerable risk. How can the risk be avoided? There is only one answer. Do the up-front work before major investments are made and, once the decision is made, monitor the total process to corroborate the original business assumptions.

As was mentioned previously, quadrants were used for simplification. The vertical and horizontal axes are really a continuum that requires a classification of new products from introducing a me-too product to managing a major breakthrough.

New Product Description in Relation to Cycle Time

Classifying or describing the different versions of new products can lead to a great deal of discussion without the resolution of the issue to anyone's satisfaction. Organizations must determine what fits their needs best. Meyer and Roberts [5] suggest a classification based on changes in technology on a four-point scale:

1. Minor improvements
2. Major enhancements
3. New related products
4. New unrelated products

This classification may serve the purpose for some organizations. It also can be reduced to only two classifications (new related products and new unrelated products), each of which could include minor improvements or major enhancements. The distinction between *related* and *unrelated* generally does not pose any problems. The distinction between *minor* and *major* must be clearly described for each business unit.

A method of classification that has proved effective for the author is a more comprehensive list that includes more classes but takes the technologies and the markets into account. New products of any type usually include technologies and markets and must be treated as units rather than separately. The classification is:

- Me-too product/service
- Improvement to a current product/service or class of products
- New product that expands the current product/service line
- Novel replacement product/service

- New-to-the-market product/service—something not available
- Breakthrough product/service

The classification "related and unrelated" suggested by Meyer and Roberts applies to the above six classifications also. Making a distinction between the related and unrelated allows management to determine just how far afield it may be going by pursuing new technological and marketing opportunities and the relations to the core competencies of the organization.

Me-too product/service

A me-too product/service is just what the name implies: a product currently marketed by other organizations. That product can utilize the same technologies, serve the same markets, and provide essentially the same features and benefits. Differentiation of competitive products in realistic terms is difficult. Generally, the organization that uses that strategy deals in commodity products and survives by being the lowest-cost supplier. Its products can be purchased from any number of suppliers.

Improvement to a current product/service class of products

Just how much effort is directed nationally to improving current products cannot be determined with any degree of certainty. It will vary from industry to industry and from company to company. The automobile industry fits that category, as does a large percentage of the consumer electronics business. Television screens may be larger, but the same operating principles apply. More gadgetry may be added to program VCR's, but not major breakthroughs. Boom boxes may be larger or smaller and have improved audio response, but the technology is essentially the same. A continuous stream of modifications that add features dominate this class of products that meet different needs. Television sets come in many different sizes with the minimum of as well as the ultimate in controls. Likewise, boom boxes come in assorted sizes, shapes, and decibel levels to fit the consumer needs. The same conditions apply to the industrial market, food services industries, and so on. Even Campbell's soups can be classified as this improvement of current products.

New product that expands the current product line

The new product that expands the current product line provides a business with many opportunities to capitalize on its knowledge of the business-related technologies and to some degree on its knowledge of the market. Investing in new technology has become a daring and expensive pastime for American business, so it is necessary to utilize the known technologies in different ways to fully exploit the original investment. That is not to suggest that a business should curtail its investment in developing new technologies. On the contrary, that investment must continue if the business is to be a viable entity in the market. The photographic and magnetic media industries provide excellent examples. The development of color film with ASA ratings of 200, 400, 1000, and so on, for the amateur market gave the user greater flexibility and capability to capture those moments of truth that previously could not take their places in the family photo treasury. The same kind of situation occurred with the introduction of new slide films and inexpensive point-and-shoot cameras. The magnetic media industry, which includes the many tapes, disks, and cassettes, exemplifies how technologies can proliferate into new products as well as new market segments.

Novel replacement product/service

Me-too products, improved products, and new products that extend the current line of products represent the major emphasis of most organizations. They are necessary, but they will not provide for continuous growth and sustained performance. They will not allow an organization to grow much beyond the growth of the economy. They will not expand markets in a shrinking economy. The mix in every business product inventory must be managed. With the novel replacement product/service classification, businesses enter a new milieu. A novel replacement product or service provides measurable important added features, advantages, and benefits to the user. The word "novel" implies uniqueness, originality, and imagination. The current audio cassette format, which has been in use for more than two decades, replaced many different configurations of audio cassettes that were not interchangeable. It also replaced the use of reel-to-reel tape. The predecessors were cumbersome, and they often consumed the tape in such a way as to make it unusable. Playback equipment was cumbersome and was also unreliable because of its complexity. The current audio cassette was certainly unique and replaced the many more complex and unreliable configurations. Other industries have seen major changes because of the introduction of novel products that replaced old methods of performing certain operations. Linotype machines are almost nonexistent; desktop publishing systems have replaced much of the manual work required in page layout and composition; xerography replaced carbon paper; and digital technology replaced much of analog technology. At a lower technological level, the paperboy running down the street shouting "Extra! Extra! Read all about it!" has been replaced by new communication technologies. The iceman was replaced by the refrigerator; the milkman no longer makes his daily rounds. Those activities were replaced by unique products or services that provided advantages to the user.

New-to-the-market products—something not available

New-to-the-market products of significance seldom appear in the marketplace. The distinction from other classes lies in the fact that such products provide something previously unavailable. The category probably receives the least consideration by management because of the risks involved. The risks and uncertainties apply equally to the development of the related technologies and the subsequent acceptance of the product in the marketplace. The distinction between novel replacement products and new to the market should be understood. The novel replacement product is a substitute for a current product with the same function. The new-to-the-market product is just that: It currently does not exist. An example is the optical disk, whether for audio, video, information, or archival storage. To this time it has not replaced magnetic media, although it provides the same capabilities. Long-play records (LPs) were once a new-to-the-market product. They replaced the 78-rpm records because they not only extended the playing time but also offered a completely new recording quality. An argument could be made that LPs were only a unique replacement product, but because of the benefits associated with their use, they more appropriately fit the new-to-the-market classification. 3M's Post-it notes are another example. Although not a very sophisticated product and not high-tech as far as the use is concerned, it requires high-tech chemistry and production capability. Computer-aided tomography (CAT) scanners, nuclear magnetic resonance (NMR) scanners, the Apple computer, fax machines, McDonald's Golden Arches, various lawn and garden services, and so on were new-to-the-market products and services.

Breakthrough products

Breakthroughs seldom occur. Although organizations search for those market, technological, and product breakthroughs, and often invest a significant amount of their resources in pursuit of them, their probability of achieving one is seldom realized. "Breakthrough" means coming up with something new, something that has not been achieved in the past. A breakthrough product involves technology and markets. In recent years it has been difficult to identify breakthrough products, technologies, or markets. Products such as the Nintendo games or

Mazda's Miata MX-5 two-seat convertible, or Gillette's no-nicks razor called the Sensor, as listed in *Fortune*'s [6] collection of the trendiest and most innovative products for 1989, cannot be classed as breakthrough products. Some innovative concepts have been used, but the products are basically of the improvement or replacement type.

Technological, product, and market breakthroughs generate new industries. Nayak and Ketteringham, in *Breakthroughs* [7] focus attention on product breakthroughs. They describe a breakthrough as:

> the act of doing something so different that it cannot be compared to any existing practices or perception. Breakthroughs, whether in commercial enterprise, science, or politics, are moments in history. Innovations are events behind the scenes that set the stage for history.

This description needs one additional consideration: Breakthroughs spawn new industries. The description of a breakthrough is a matter of personal choice. Nayak and Ketteringham consider Post-it notes as a breakthrough. According to the suggested sequence of product classification it is a new-to-the-market product. Breakthrough products from the past would include products that came about as a result of breakthrough technologies. They began with the steam engine and other inventions such as the telephone, the electric light bulb, movies, radio communication, and then span the complete spectrum of more recent technologies that resulted in a proliferation of new products such as color photography, xerography, digital techniques in many different configurations, superconductivity, jet aircraft, biomedical introductions, space exploration, and mechanical and electronic medical technology. Those breakthroughs were not planned events. They emerged from the minds of individuals capable of observing a multiple number of events and then synthesizing what was learned into some cohesive conceptual framework. Where are the breakthrough technologies that will provide the industries that at some time in the future will replace digital systems, xerography, the automobile, and so on? There are no definitive answers. There is one certainty, however: They will be replaced by breakthrough products at some time in the future.

Significance of classification

Although there is no need to overemphasize new product classification, it is necessary that managers understand where they are placing the emphasis. There is no need to split hairs in these classifications. The purpose of the classification is to allow a business unit to ultimately determine where it is or where it should be investing in the continuum of new products. Figures 5.1 and 5.2 show the relations of technologies, products, and markets. The product axis in each case includes a continuum of new product descriptions. The same approach would not be used for introducing both a me-too product and a new-to-the-market product. The difficulties associated with introducing each class of product would be different and would involve acceptance of various degrees of risk. It is essential, though, that managers understand the particular milieu in which they are working. The purpose of the distinction is to convince managers to begin thinking about what benefits will be derived from introducing a new product. As an example, product improvements seldom generate large increases in market share. At best, they make it possible to maintain current market share.

Product Life Cycle

Articles in the business periodicals and the daily press continue to stress that product life cycles are shorter than in the past and therefore businesses must speed up the cycle of new product introductions. There is no doubt that product life cycles have become shorter. However, product life cycle must be considered in relation to the product changes that are taking place. As an example, the product life cycle of products embodying the technologies related to xerography can be considered very long or very short depending on the perspective from which it is viewed.

The xerographic process was invented by Chester Carlson in 1938. A copy machine was unveiled in 1948, but it was more than a decade before Haloid-Xerox would deliver a practical rotary-drum office copier. The fundamental technology goes back more than a half century. An argument could be made that indeed the life cycle of xerography is more than 50 years—not very short by any scale of measurement. Major improvements in the process have been made over the years; many important enhancements have been added. Techniques for enlarging or reducing an image, collators, and systems to help the user manage the machine operations have been added by the various manufacturers in the industry.

The short product cycle time has little to do with the fundamental technologies and concepts underlying xerography. It involves the addition of enhancements, relatively modest technical improvements, and resolving field operating problems to gain some advantage over the competition. In spite of all of the emphasis on quality by all manufacturers, including U.S. and Japanese, originals continue to be consumed by machines, copy quality does not always come up to the required standards, and certainly none of the machines pass the tests for noise elimination. Engineers may attempt to design the perfect machine, but the system will never function effectively and efficiently 100 percent of the time: People are involved; the material to be copied will span a wide range of acceptability; and the paper will most likely not have been stored under ideal conditions. Improvements in each of these areas will continue, but the products are not new. Considering each improvement as a product life cycle leads managers into thinking that they really have introduced a product with significant features, advantages, and benefits, whereas in reality the improvements are modest.

The typical life cycle curve in Fig. 5.3 shows the sales and profits through the various phases of product introduction, growth, maturity, saturation, decline, and withdrawal. Such an idealized curve can, however, lead to some false assumptions. Seldom does an organization focus total attention on a single product. It may direct attention to a single class of products that includes many different configurations and serves many different markets. Copy machines form a class of products. Regardless of speeds, sizes, and differences in complexity, they remain copy machines. A typical product line may undergo many improvements and modifications during its life cycle, but it only serves the same purpose in a different way. If as some managers insist, product life cycles are often less than 6 months, the introduction, growth, maturity, saturation, decline, and withdrawal periods take place in 6 months or less. Talking short cycle times may justify lack of performance in the marketplace, but the reality, as shown by the Sony Walkman, is the same basic product is being introduced more often with minor modifications and additions to meet individual needs more effectively. Emphasizing short life cycles generates a false impression of rapidly changing competitive markets, whereas the short cycles are of relatively minor improvements. Emphasis on short product cycle time cuts across all industries from the auto to the computer to the

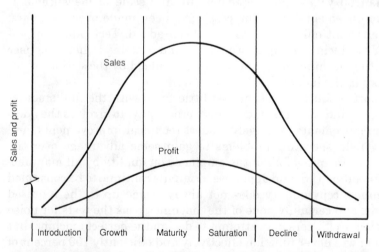

Figure 5.3 Life cycle sales and profits from introduction to withdrawal from the market.

other technology-driven industries. Product cycle time must be viewed realistically. A clear distinction must be made between the product cycle time relative to improvements or differentiation and the product life cycle of the original product and its many modifications or enhancements. Some improvements provide a measurable benefit; others are only cosmetic.

Information Transfer from Research to Marketing

Rather than consider the issues solely related to technology transfer, it is important to draw attention to total business information transfer. Technology transfer is only one element of information transfer. Although transferring technology at all levels is important, by itself it is insufficient. Figure 5.4 illustrates product information transfer from research to marketing. The vertical axis is the percent of product information available from the delineation of a concept to market acceptability and the achievement of the sales objectives. The horizontal axis, the time line, includes research, development, manufac-

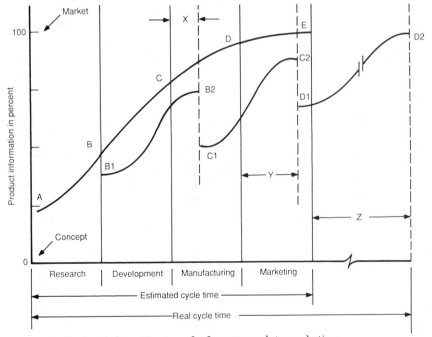

Figure 5.4 Product information transfer from research to marketing.

turing, and marketing. For purposes of simplification, each of the functions is allocated the same time.

A typical theoretical or ideal curve would start at point A, continue to points B, C, and D, and stop at E. Point A is at about 25 percent, since the assumption is that some substantive knowledge is available before the project begins. To avoid discussing the merits of different organizational structures, research, development, manufacturing, and marketing are thought of as activities that must be performed rather than specific organizational functions. Whether an organization chooses to use teams, matrix structures, or the strict adaptation of discrete business functions depends on organizational leadership.

The ideal curve from point A to point E is seldom accomplished. The primary reason can be the lack of adequate and timely information transfer. A more continuous as well as comprehensive approach to information transfer provides opportunities for approaching the results shown in the ideal curve from A to E in Fig. 5.4. The usual progression is that development, instead of beginning at point B, actually begins at point B_1. To transfer the knowledge generated by research, additional time is required, so instead of completing the development activity at point C, additional time is required and development is now completed at point B_2. The lack of continuity between research and development causes an extension of time X. Whereas manufacturing was to be at point C, in reference to product knowledge on the ideal curve, it is now delayed by time X and begins its work at point C_1. Once again a knowledge gap of the prior activities in research and development exists in manufacturing. Manufacturing now concludes its efforts at point C_2, and the delay in time to accomplishment is now represented by the letter Y.

Marketing, according to the ideal curve, should have been completing its market introduction at point D. Because of the series approach, marketing now begins at point D_1. It completes the full required knowledge at point D_2, but with a time delay indicated by the letter Z. In other words, the cycle time for this particular development was extended by time Z. In this case there is a significant disparity between what is shown as the theoretical or projected cycle time and the real cycle time. Delays of this type can be traced to lack of adequate information transfer. Figure 5.4 represents the typical series approach to product development and market introduction that significantly extends cycle time. Simultaneous engineering can, at best, affect only the transfer of information from development to manufacturing. Multifunctional teams may or may not be the answer; if they are, they must meet the requirements of the ideal curve that begins at point A and proceeds to points B, C, D, and E.

Interaction of Research, Development, Manufacturing, and Distribution

Figure 5.5 illustrates the various levels of involvement by research, development, manufacturing, and distribution in optimizing cycle time. Distribution includes marketing and sales, physical distribution, and customer service, but it will be considered primarily from a viewpoint of marketing and sales. That format follows the basic organizational

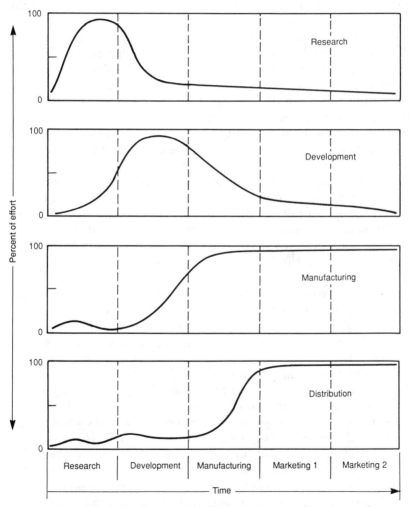

Figure 5.5 Percent of effort by research, development, manufacturing, and marketing people during the phases normally described as research, development, manufacturing, and marketing.

structure shown in Fig. 2.2, the tripartite organization. The administrative entity is not included in this example, but it must be taken into consideration because it will affect the product cycle time. The curves in Fig. 5.5 represent the involvement by the various disciplines and not the functional departments.

During the research stage, specialists in the related research will undoubtedly be the most active participants. In the early stages, though, marketing must provide sufficient input and effort to assure that the directions taken by the researchers will meet the market requirements. Research as used in this context involves the resolution of the technology issues related to a specific product for which no apparent solution exists. The solutions require more than the application of off-the-shelf technology packages. During this stage, development and manufacturing must become informed about the details of the research in order to plan their own activities and also to guide the researchers toward solutions that can be implemented within the scope of the business unit resources. It is at this stage that decisions may have to be reached regarding the availability of plant and equipment, adequate processes, raw materials or substitutes, and so on. The distribution functions also must consider the marketing and sales strategies, the adequacy of the physical distribution systems, and the issues related to customer education and product performance. This period gives the researchers an opportunity to determine if more appropriate solutions would be possible if new capital equipment were provided. Could research time be shortened if some new process that could be justified within the scope and the financial limitations of the program were developed?

During the development stage, research continues to be involved, although to a much lesser degree. Development includes design, prototype or pilot plant operations, and product testing. It includes all the elements that may be required to demonstrate the capability of a product. It involves solving all the product problems before going to the factory floor. The marketing specialists, though, need another major effort with the development people to guarantee that the integrity of the product specification and system, as conceived and perceived at this time, is maintained. Changes in direction by individuals without concurrence by the other functions cannot be tolerated. This period often resembles the situation that occurs when a married couple decide to build a new home. The plans are completed and approved. The quotations have been received and evaluated. The contractor is selected, and the contracts are issued. After construction begins, changes are requested without any consideration of the implications of the changes not only on added cost and completion schedule but on the functionality of the building. The 1950s movie *Mr. Blanding Builds His Dream House* demonstrated the situation. Unfortunately, too many executives and managers follow Blanding's philosophy in managing the introduction of new products. Changes are made without effectively communicating the changes to all the participants (internal as well as external) and without full consideration of their impact on the product cycle time, the timing to the marketplace, and total time.

During the manufacturing stage, research continues to play a minor role, development continues to be involved to assure that the designs will be manufactured as developed, manufacturing becomes a full-time participant, and marketing begins to increase its activity according to the needs.

Marketing is split into two stages: Marketing 1 involves introducing the product to the market and racking up the sales to meet the original projections. Manufacturing continues to be a full-time participant; research and development reduce their effort but continue to provide input as required. But research and development need feedback from manufacturing and marketing. That feedback is absolutely essential. It is at this stage that much can be learned about product performance and the future needs of the customer. It is an opportune time to assess what, if done differently, would have yielded better results. This is not a time for manufacturing to make changes without the involvement of those who developed the product. That would be product suicide. Marketing 2 represents the time for capitalizing on the investment. This period must be supported by specialists from other functions as required. The specialists should, however, be aware of problems occurring in the field. That information is generally ignored in most U.S. businesses. In my contact with several Japanese organizations, I found that weekly customer feedback to research, development, and manufacturing was part of the management process. That information was not restricted to marketing and sales.

The process from product concept to commercialization is continuous. The improvements to the first entry of a product cannot be delayed until the product reaches maturity. The current competitive market requires, in many instances, that work on the second or improved entry commence before the first has left research or development. The relations of the business functions shown in Fig. 5.5 demonstrate that in reality there should be no need for formal information transfer from one function to another. All the functions should be aware of the activities in the other functions and how they affect their particular activities. The current approach used by businesses focuses on the use of teams. Although teams provide one means of improving information transfer, it is necessary to understand that how those teams are organized and managed ultimately determines whether the three elements of cycle time will be effectively satisfied.

Smith and Reinertsen, in *Developing Products in Half the Time* [8], discuss the use of overlapping activities as suggested in Fig. 5.5 in relation to information transfer. They consider the issues related to degrees of overlap for reducing cycle time. They state that the part-time phase-in concept encourages a sequential approach, and they therefore recommend full-time staffing from the various disciplines. However, project staffing does not depend on disciplines; it depends on specific knowledge, experience, and skills. Certain individuals may be required full time on a project, but many of them will be selected solely for their specific expertise. As an example, manufacturing is not required to participate full time in the early stages of research and development. Full-time manufacturing participation from the beginning will only add unnecessary hours to the project. If on a specific project manufacturing can make a sustaining contribution to the project from the beginning, then the appropriate decision would be to include manufacturing full time. That is true of marketing and other functions also.

Team or task force management requires that overlapping of functional talent and cross-functional interaction at all levels be controlled. People are a resource, and to allow them to vegetate in meetings or some form of team structure in which they participate at some fraction of their competence level is management malpractice.

The set of curves in Fig. 5.5 and the relative involvement of the functional specialists depends on the type of product. Every situation will require different levels of involvement. The curves would be quite different depending on where the product lies on the spectrum from a me-too product to a breakthrough product. A new-to-the-market or a breakthrough product will require much more time for research and development than a product improvement. A research phase may be lengthy and span many years, but that does not allow the other functions to ignore the research. It is absolutely essential that they offer their input in relation to the approaches taken by research.

Most research in product development focuses attention on discrete parts manufacturing. The parts, made from many different materials, and some electrical and electronic subassemblies are configured into a product that meets certain specifications. The process industries also must be considered. Discrete parts manufacturing and the process industries differ considerably in their approach to new products. Process industries usually require more long-range thinking. Arranging some machine tools into networks to function more efficiently is quite different from designing new paper mills or various types of coating machines such as those used in producing photographic film and magnetic media. A similar statement applies to the fine-chemical and petroleum-related industries. The lead time in those industries to bring a system on stream and then scale up many different products with different operating parameters may be three or more years.

New Product Process Assessment

Optimizing product cycle time requires a process. There are no step-by-step recipes that can be followed, since the process must meet the requirements of the specific business unit. But some general principles can be provided, and certain conditions must be met in order to assure an acceptable level of success. Research is very limited in this area of utilizing formal processes for introducing new products. Much of it is anecdotal, but that is not a reason to ignore the lessons that can be learned from such examples even though related to a single business. Although researchers frown on and discredit the value of anecdotal information, they forget that all they really accomplish in their research is the averaging of many anecdotes. Formal studies are further flawed by the fact that they usually involve information from top-level executives and CEOs who do not necessarily know what is actually taking place in their organizations. They know what should take place, but not what is actually taking place. The business press demonstrates that with every publication of business news. Briefing papers from staff specialists and lower-level management seldom tell it the way it is. In any case, managers can learn from the experiences of others.

According to Booz-Allen and Hamilton, some firms have implemented a systematic process for moving a product from concept to commercialization [9]. My own research in this area indicates that to greater or lesser degrees successful organizations at least have a detailed process for new product introductions, although it may not be followed rigorously and not be used in many other situations. The processes continue to focus on the series approach going from research to development to manufacturing to marketing. Steps that are costly to correct or lead to unsuccessful results are often omitted. Whether formal processes provide for more timely introduction of products has not been proved by any research. This question might be asked: Is it necessary? Quantitative verification would be difficult to compile. The determining criteria of what constitutes *meeting the requirements* are not fixed. Each situation must be evaluated on the basis of the particular circumstances. In some case, the process may not be totally delineated. It may have become routine; it may be embedded and accepted in the business culture. Whether that approach is viable over a long time period is questionable.

Booz-Allen and Hamilton estimate that during the period 1981–1986 new-product introductions doubled. They also state that by the end of 1986 the new products introduced in that 5-year period would account for 40 percent of corporate sales. Such figures must be viewed with some discretion; the question is how a "new product" is described. At best, most new products were improvements of existing products. It is difficult to find many new-to-the-market or breakthrough products during this period.

There are no mysteries in developing a new-product process. It requires common sense, intuition, a proper balance of qualitative and quantitative data, a logical parallel/series sequence, and consideration of alternatives for potential uncontrollable events. Every organization needs to take a realistic look at its new product process. The facts must be brought out on the table and discussed openly in order to resolve the conflicting opinions and separate them from the facts.

Reviewing the research literature for why products fail or what conditions are required for success provides some insight into the essential elements of the process. The results of a study by Cooper [10] are given in Fig. 5.6. The figure shows the 13 key activities identified by Cooper that are involved in the new product process. One could argue that the list is insufficient or too extensive. The fact is that the manner in which each activity is performed determines the product cycle time to some degree.

New Product Process Activities

- Initial screening process
- Preliminary market assessment
- Preliminary technical assessment
- Detailed market study/market research
- Business financial analysis
- Product development
- In-house product testing
- Customer tests of product
- Test market and trial sell
- Trial production
- Precommercialization business analysis
- Production start-up
- Market launch

Conclusions of Cooper Study

- No market study or detailed market research was conducted in three-quarters of the projects.
- 77 percent of the projects did not involve a market test or trial sell.
- 34 percent omitted product testing with the customer.
- 32 percent did not use a formal product launch.
- 65 percent lacked precommercialization business analysis.
- 37 percent omitted the business and financial analysis prior to product development.

Figure 5.6 Results of a study by Cooper: the 13 new-product process activities and the conclusions.

Cooper's study of 203 new-product projects also revealed that the complete set of 13 new-product process activities, as defined, were followed in only 4 of the 203 projects. Successful products followed a more disciplined approach and included a preliminary market assessment, a precommercialization business analysis, and a formal market launch in 20 percent more successes than failures. Also, technical assessment was undertaken in 14 percent more of the successes than failures.

The relative importance of each of the 13 activities in the product process depends on the purposes of the organization and the products or services of the business. Cooper notes that successful products were characterized by much better execution, but that is self-evident to those involved in product development. The 13 factors should certainly be a part of any organization's product introduction discipline.

Cooper does not provide any details regarding the types of products that resulted from the 203 projects. The assumption is that they were products of some type for the commercial or consumer market and did not involve a significant investment in new or upgraded capital equipment. In other words, Cooper focused on products that result from an assembly of discrete parts, mechanical or electronic, and not on the process industries. Another assumption is that the products considered were small and would exclude projects related to development of new aircraft, total information systems, automation, and so on. Projects of large scope are not afforded the luxury of prototypes, production start-ups, test marketing, and so on. As an example, introducing a new aircraft does not allow the first customer to totally evaluate the aircraft prior to a commitment to purchase. An organization that designs and builds specialized manufacturing equipment at a cost of many millions of dollars for a single unit is not allowed the luxury of building prototype units for test purposes. Such a project, even though part of a production process, can be classed as a research and development project. The prototype, or first unit, is the new production plant. An opportunity to duplicate the latest seldom occurs.

In a related article, Cooper and Kleinschmidt [11] reviewed the literature regarding the successes and failures in new products in an effort to develop a set of hypotheses about successful products. The objective was to validate the hypotheses. From the many publications they compiled a list of the activities necessary for product success. Their conclusions of the study are given in Fig. 5.7.

Cooper and Kleinschmidt have taken the results of the empirical investigations of new-product introductions, which are consistent in their conclusions, and postulated 10 hypotheses, 9 of them related positively and 1 negatively. Figure 5.8 lists the results of their study. The figure lists the 10 hypotheses, the measures for success, and the correlation between the hypotheses and the measures for success. This study, like

Recognition of Technical Opportunity

- Market need recognition
- Proficient R&D management
- Ample development resources
- Close link to market needs
- Market research
- Information acquisition
- Collaboration between innovator and user
- Strong internal communication
- Highly developed screening procedure
- Good company product fit
- Utilization of technical know-how
- Early market need recognition
- Product champion

Causes of Failure of New Products
- Inadequate market analysis
- Product defects
- Lack of adequate marketing effort
- High costs
- Bad timing
- No test marketing
- No financial evaluation

Conclusions of Project SAPPHO
- Understanding the user's needs
- Attention to marketing and publicity
- Efficiency of development
- Effective use of outside technology and external scientific communication
- Seniority and authority of responsible managers

Conclusions of Project NewProd
- A unique and superior product in the eyes of the customer, one with a distinct and important advantage
- Good knowledge of the market and the market inputs and being proficient in market research and the marketing tasks
- Technological and production synergy and proficiency

Figure 5.7 The conclusions of the Cooper and Kleinschmidt study of activities related to product success.

Ten Postulated Hypotheses

- Product advantage
- Market potential
- Market competitiveness
- Marketing synergy
- Technological synergy
- Project definition or "protocol"
- Proficiency of the up-front activities
- Proficiency in market-related activities
- Proficiency in technological (development and production) activities
- Top-management support

Measures of Success

1. Success or failure: Did it or did it not meet the acceptable profitability level?
2. Profitability level: the degree to which the product's profitability exceeded or fell short of the minimum acceptable level.
3. Payback period: the number of years to recoup the investment.
4. Domestic market share: domestic market share in the third year after product launch.
5. Foreign market share: foreign market share in the third year after project launch.
6. Relative sales: dollar sales of product relative to recent new product introductions.
7. Relative profits: profit level relative to recent product introductions.
8. Sales versus objective: extent to which sales exceeded or fell short of objectives.
9. Profits versus objective: extent to which profits exceeded or fell short of objective.
10. Opportunity window on new categories: extent to which investment yielded a new class of products.
11. Opportunity window on new markets: extent to which product allowed entry into new markets.

Correlation Between Hypotheses and Measures of Success

- Product advantage—correlated with 10
- Market potential—correlated with all 11
- Market competitiveness—no correlation
- Marketing synergy—significantly correlated with 6
- Technological synergy—significantly correlated with 7
- Project definition or "protocol"—correlated with 8
- Proficiency of the "up-front" activities—correlated with all 11
- Proficiency in market-related activities—correlated with 7
- Proficiency in technological activities—correlated with 8
- Top-management support—low correlation coefficient with 5

Figure 5.8 The results of empirical investigations by Cooper and Kleinschmidt into new-product introductions.

all studies, is limited in scope but provides direction. One can argue with the validity of the hypotheses and the measurement criteria. The academic researchers can make their quantitative analyses, but the fact remains that all the hypotheses to some degree or other affect new product introduction into the marketplace. I seriously question the relative difference between a correlation with 11 of the criteria or 7 or 8 criteria. That difference depends upon the product.

Proficiency in doing the up-front work correlates with all 11 measures of success. That is and will be a theme throughout this book. The up-front work determines level of success because, to be effective and meaningful, it must take into account the business as a system and recognize that the resources and the infrastructure must be in place. Another interesting result of the Cooper-Kleinschmidt study is that, contrary to much of current thinking, top-management support is the least important. That does not mean it is not important at all. This result must be taken into consideration as researchers continue to stress that top-management support is essential for introducing new approaches to management, new technologies, creating organizational change, and so on. Granted that top-management support can ease the way to accomplishment. But the individual who chooses to create something new, if self-motivated and equipped with the drive for accomplishment, will discount the odds and pursue his or her particular objective with or without that support.

In another assessment related to introducing new products, Gupta and Wilemon, in "Accelerating the Development of Technology Based-Products" [12], examined the attitudes of managers and professionals related to:

- Reasons for accelerating product development
- Reasons for product development delays
- Major concerns of team members
- How functional groups delay the process

Figure 5.9 provides the details under each of those headings.

The reasons stated by the managers and the professionals for accelerating the product development process appear to be rather myopic. Managers are reacting to the environment rather than trying to create it. All the reasons stated have validity, but one would think that, in this day of sophisticated technology and information systems, more relevant reasons might emerge. There is no doubt that the reasons stated are valid, but they are insufficient as to why product development cycle time should be accelerated. Perhaps the results substantiate the change in mindset that has taken place in industry in the last two decades. Competition was considered as the major reason for accelerating product development. Shortening not only cycle time but the total

Reasons for Accelerating Product Development
- Increased competition (42%)
- Rapid technological changes (29%)
- Market demand (11%)
- To meet growth objectives (11%)
- Shortening product life cycle (8%)
- Senior management pressure (8%)
- Emergence of new markets (5%)

Reasons for Product Development Delays
- Poor definition of product requirements (71%)
- Technological uncertainty (58%)
- Lack of senior management support (42%)
- Lack of resources (42%)
- Poor project management (29%)
- Other (20%)

Major Concerns of Team Members
- Management style (53%)
- Lack of attention to details (47%)
- Limited support for innovation (32%)
- Lack of strategic thinking (18%)
- Poor manufacturing facilities (16%)

How Functional Groups Delay the Process
- Failure to give program priority (58%)
- Continually changing requirements (58%)
- Poor intergroup relations (34%)
- Slow response (26%)

Figure 5.9 Results of Gupta and Wilemon study about the attitudes of managers and professionals in accelerating new product development activities.

time required from concept to market introduction is not only a matter related to the competition. It would appear that improvement, whether immediately required or not, would be a motivating force for managers and professionals to push toward new horizons. The response to the Gupta and Wilemon questions seems to be that unless someone is going to force an organization to improve, nothing will be attempted. The proactive element is totally lacking.

The responses to the new-product development delays includes the usual reasons. Whether the percentages in such research are significant depends on the business culture in which the individuals work and the level of self-justification in nonperformance. The question of lack of senior management support contradicts to some degree the results of Cooper and Kleinschmidt that indicated that management support had little impact on determining the success of a product. If major investments of time and money are required, some support must be present, although it may not be exhibited by continued personal interest. It depends on how management support is described.

Management style, a somewhat undefinable characteristic, is the major concern of the team members and receives the greatest attention. Without doubt, management style may affect new product development. But is it style or culture? If it is related solely to style, it may only indicate that the style that managers use is not the preferred one. Just what a respondent means by "style" must be clearly delineated. Culture to a great extent determines management style. The high percentage allocated to the lack of attention to details also raises some questions about what details and by whom. In each of these four sections of reasons for delay of new-product introductions, the emphasis or the reasons for extended cycle time appear to be directed at *some other person*. It does not appear that the participants in the study consider *their own limitations and lack of performance*—someone else should be proactive.

It is interesting to note that failure to give program priority and changing requirements are at the top of the list of how functional groups delay the process and extend cycle time to market introduction. The results clearly show that functional objectives take precedence over business objectives or that business objectives are not communicated effectively. It also points at the management malpractices current in many organizations. As noted previously, agreement on product specifications is a necessity, and it must be continuously monitored. That is not a matter of choice. Poor intergroup relations point not only to management but to the participants. In the final analysis, they determine the intergroup relations. The participants also determine the speed of response in their own activities. It is not only a matter of what others do or do not do. The team has the responsibility for resolving those issues.

Gold, in "Approaches to Accelerating Product and Process Development" [13], suggests three strategies for accelerating product and process development:

- Use of external resources
- Intensified internal research and development (R&D)
- Promoting innovative R&D management strategies

Peer review to accelerate progress

Although there is no argument with the three strategies suggested by Gold, experience in dealing with product development activities clearly shows the absolute necessity for integrating the technology functions of an organization and treating the business as a system. Research and development are not individual entities; they are integral parts of the business system. That is the point that is missing. Research shows that technology can be developed just as successfully internally as externally. Its success depends on the technology and the attitude with which the participants approach joint development. Innovation in managing research and development is absolutely essential, but it will not come about by introducing only the latest approaches to managing teams, empowerment, or self-managing groups. Innovation in R&D management requires leadership in creating change. It requires ideas—ideas that can be implemented with the available resources and the business infrastructure. Innovative management strategies must go beyond such simple approaches as peer review, transfer of responsibility, balanced R&D portfolios, and closer integration of R&D with other functions, as suggested by Gold. New ideas are not usually accepted by the ruling bureaucracy even though that bureaucracy consists of skilled scientists, engineers, and marketers.

Dwyer and Mellor, in "New Product Process Activities and Project Outcomes" [14], review the research of Cooper, Kleinschmidt, Booz-Allen and Hamilton, Burgleman, and others and compare the results of their own research in the United Kingdom, Belgium, and Australia. Their findings essentially corroborate those of past researchers. It is difficult to make exact comparisons of studies by different researchers, but the findings are consistent. Their study once again emphasizes the need for effective performance of the up-front or predevelopment activities and the need for a disciplined new-product process.

Ancillary Considerations

Success in optimizing product cycle time depends on whether the organization takes a reactive or proactive approach. An organization can accept the fact that delays will occur, or it can take a proactive approach and resolve the issues that extend cycle time beyond acceptable limits. It can focus on time allocation in the individual functions, or it can address the issues related to the three elements of cycle time: time, timing, and cycle duration. The organization can pursue the problems from a holistic perspective or limit it to the interfaces associated with simultaneous engineering. One point must be clearly recognized: product and process are one; they are not independent [15]. Optimizing product cycle time challenges the total organization.

The literature related to product development contains many hypotheses that are not substantiated in practice. The idea that product and process development are mainly a creative data gathering and processing exercise loses sight of need to translate the data into some physical embodiment [16]. Generating new ideas requires more than gathering and processing data. There must be a synthesis of information and knowledge that flows from processing the data. Something must be demonstrated. The use of brainstorming sessions by inexperienced people that produce exciting ideas will not yield a continuous flow of new products to the marketplace. If the process is so simple, why are new-to-the-market products and breakthrough products so difficult to identify?

De Meyer and Van Hooland, in "The Contribution of Manufacturing to Shorten Design Cycle Times" [17], conclude that improvement in new-product introduction time does not require significantly greater amounts of financial resources in research and development. Throwing money at the problem is not the answer. Developing major new programs in reducing product cycle time is not the answer. But commitment of the total organization is essential. All of the administrative functions must also become proactive.

Procurement strategies play a major role in managing new product cycle time. Morgan, in "Are You Aggressive Enough for the 1990's?" [18], reports on the research of Robert Monczka, professor of purchasing and materials management at Michigan State's Graduate School of Business Administration. Monczka's study shows that, although continued emphasis will be placed on gaining market share, improving return on investment, and achieving low-cost producer status, emphasis will also be directed at developing the ability to move from concept to market in a shorter period of time. He emphasizes that joint buyer/seller efforts on supplier cost will be needed to maintain competitiveness. Those efforts will also have to be directed toward improved materials with physical characteristics related to the needs. Monczka is one of few academics who recognize the need for taking the system approach in introducing new products. He states that purchasing must be integrated with product strategy (normally it is disregarded), but he also notes that purchasing managers must get their act together. Purchasing must move away from the traditional approach of buying parts and raw materials. The new breed of purchasers that will have to be developed will "think like executive management." In other words, they are participants in the business and must become knowledgeable beyond their limited functions.

Summary

This chapter has focused attention specifically on the factors related to product cycle time. The purpose is to consider the conditions that must prevail if businesses are to make a significant improvement.

- Product cycle time must be considered from the time a concept is delineated until the embodiment has been replaced or displaced in the market.

- A strategy for product cycle time must be developed. As in any business endeavor, that strategy must flow from purposes and objectives.

- Product cycle time must be considered in relation to product classification. It will be quite different for an improvement to a current product as contrasted to introducing a new-to-the-market product, and so on.

- A clear distinction must be made between product cycle time and product life cycle. Product life cycle must be put into a business perspective. The differences between improvements in the cycle and replacement or displacement must be recognized.

- Technologies, products, and markets must function as unities. They interact and depend on their level of integration in affecting product cycle time.

- Optimizing product cycle time depends on transfer of information and not only on transfer of technology. New product introduction involves the complete organization. The validity of the information that is transferred must at all times be questioned to ascertain its applicability.

- Research, development, manufacturing, and distribution (marketing and sales) are one and must be treated holistically. Figure 5.5 shows the continuous involvement of each function to the extent required throughout the lifetime of the product.

- Product cycle time and success depend on introducing new products that offer product advantage, market potential, marketing synergy, technological synergy, product definition, proficiency of the up-front activities, proficiency in market-related activities, and proficiency in technological activities. It also includes integration of people and organizational processes, of managers and participants.

- Freedom is an essential ingredient for any organization, but it must be disciplined. Managers and professionals must recognize that they are not free to do whatever they choose.

References

1. J. H. Sheridan, "Racing against Time," *Industry Week,* June 17, 1991, p. 23.
2. J. L. Bower and T. M. Hout, "Fast-Cycle Capability for Competitive Power," *Harvard Bus. Rev.,* November-December 1988, pp. 110–118.
3. A. Rosenblatt and G. F. Watson, "Concurrent Engineering," *IEEE Spectrum,* July 1991, pp. 22–42.
4. "Products: Variations on a Theme," *Harvard Bus. Rev.,* January-February 1991, pp. 112–117.
5. M. H. Meyer and E. B. Roberts, "Focusing Product Technology on Corporate Growth," *Sloan Manage. Rev.,* 29(4), Summer 1989, pp. 7,8.
6. E. C. Baig, "Products of the Year," *Fortune,* Dec. 4, 1989, pp. 162–170.
7. P. Nayak and J. M. Ketteringham, *Breakthroughs!,* Rawson Associates, New York, 1986, pp. 50–74 and 343–347.
8. P. G. Smith and D. G. Reinertsen, *Developing Products in Half the Time,* Van Nostrand Reinhold, New York, 1991, pp. 153–168.
9. Booz-Allen and Hamilton, *New Product Management for the 1980's,* Booz-Allen and Hamilton Inc., New York, 1982.
10. R. G. Cooper, "The New Product Process: A Decision Guide for Management," *Journal of Marketing Management,* vol. 3, no. 3, Spring 1988.
11. R. G. Cooper and E. J. Kleinschmidt, "New Products: What Separates Winners from Losers?," *Journal of Product Innovation Management,* vol. 4, no. 3, September 1987, pp. 169–184.
12. A. K. Gupta and D. L. Wilemon, "Accelerating the Development of Technology-Based Products," *California Management Review,* vol. 32, no. 2, Winter 1990, pp. 24–44.
13. B. Gold, "Approaches to Accelerating Product and Process Development," *Journal of Product Innovation Management,* vol. 4, no. 2, June 1987, pp. 81–88.
14. L. Dwyer and R. Mellor, "New Product Process Activities and Project Outcomes," *R&D Management,* vol. 21, no. 1, 1991, pp. 31–42.
15. G. H. Gaynor, *Achieving the Competitive Edge through Integrated Technology Management,* McGraw-Hill, New York, 1990.
16. J. M. Utterback, "Innovation in Industry and Diffusion of Technology," in M. Tushman and W. L. Moore, *Readings in the Management of Innovation,* Pitman Books Ltd., Marshfield, MA, 1982.
17. A. De Meyer and B. Van Hooland, "The Contribution of Manufacturing to Shortening Design Cycle Times," *R&D Management,* vol. 20, no. 3, 1990, pp. 229–239.
18. J. P. Morgan, "Are You Aggressive Enough for the 1990's?" *Purchasing,* April 6, 1990, pp. 50–57.

Project Cycle Time

Project cycle time and project management determine business performance. Archibald [1] states that a firm's strategy is implemented through projects. Project cycle time receives the most attention from executives and managers. Although it may have been delayed years, once the decision to approve a project has been made, speed seems to take precedence over all other considerations. Whether the up-front work was executed diligently or haphazardly makes little difference. It is time to get moving, release those purchase orders, start cutting metal, and so on, even though product specifications have not been finalized and future market requirements have not been evaluated. It may have taken a long time to reach the decision to proceed, but the activities related to project cycle time are squeezed to the point at which there is never sufficient time to do the right thing the right way but always time to do it over and over and over. This chapter will not deal with the usual topics associated with project management such as planning, scheduling techniques, and methods. It emphasizes the aspects of project cycle time that can provide the innovative spark that leads to improved business performance. That means thinking and doing things differently.

Too often, executives and managers, including project managers, take a very narrow view of just what is involved in project cycle time. A systems approach to project cycle time takes into account not only the details related to the management of the project from the time of formal approval but also the activities and decisions that preceded the authorization. Figure 6.1 relates the continuum of cycle time from concept to commercialization. The project activities involved in developing a new-to-the-market product will clearly demonstrate the major causes for extended project cycle time. This chapter considers the primary issues related to the following:

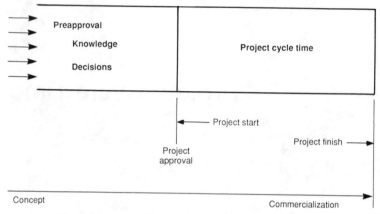

Figure 6.1 The continuum from concept to implementation.

- Concept of the project
- Description of project cycle time
- Integrated approach to managing project cycle time
- Strategic and operational issues
- Linkage of technology, marketing, new products, and cycle time
- Key issues in optimizing project cycle time
- Lessons from a case study
- Managing the enterprise with the project approach

Concept of the Project

What is a project? A project can be described as including a set of specific activities of a group of people organized and dedicated to achieving an objective. In essence, it involves the allocation and use of the ten business resources in pursuit of a specific objective. There are many different types of projects: those related to the technology functions that involve both products and processes, capital projects related to providing new factories, upgrading production equipment, building water purification systems, and so forth, and environmental and safety projects. Projects come in all sizes from less than a one-person effort to hundreds of people from many different disciplines. The cost can vary from practically zero to megamillions. Examples range from designing, building, and operating the tunnel under the English channel to making a preliminary study of some minor product improvement. Both are projects, although at the extremes of the spectrum. The tunnel is an example of a system involving many decisions and many people crossing national boundaries; the minor product improvement represents an individual effort with different degrees of involvement of other professionals. There may be larger projects than the connecting tunnel, and there certainly are many projects smaller than the study. In between the extremes lies an assortment of projects related to research, development, manufacturing, and marketing of many products and services in different industries. In the final analysis all are projects.

Ritz [2] relates the following features common to managing capital projects; actually, they apply to any project:

- Each project is unique and not repetitious.

- A project works against schedules and budgets to produce a specific result.

- The project team cuts across many organizational and functional lines that involve virtually every department in the company.

- Projects come in various shapes, sizes, and complexity.

Although Ritz discusses project life cycles, he excludes them when he considers the common features. Project life cycle belongs in the list of those common project features. Whether Ritz's specific classification is used is not important. Every organization must describe the cycle as it applies to its business.

The project approach provides an orderly process for accomplishing some specific objective. It takes into account all the activities and decisions in meeting those objectives. The degree of success spans the continuum from totally unsuccessful to achieving more than was expected. In the final analysis, the level of success depends on people and interactions with the seven elements of the business infrastructure: purposes, objectives, strategies, organizational structure, guiding principles, policies and practices, and management attitude. That orderly process, which is iterative and evolves as a project progresses, although not conforming to a specific recipe, can be guided by some sound principles. Figure 6.2 lists some of the requirements for successful new-product projects as discussed in Gaynor, *Achieving the Competitive Edge through Integrated Technology Management* [3].

Description of Project Cycle Time

Project cycle time begins with authorization and ends when total implementation has been accomplished. It includes the time duration the total time in hours required to achieve the stated objectives as well as the proper timing. It includes all the delays that occur because of changing requirements, reconsideration of initial objectives, availability of new

1. *Understanding the business, the technologies, and the markets.* Business includes the infrastructure and the resources; technologies include the knowledge that can be derived from a comprehensive *business unit technology plan* and the marketing and sales knowledge as it applies to the project under consideration and the future directions of the organization.

2. *Clearly stated, articulated, and understood project objectives.* Although there may be a single primary objective such as "introduce product X at some specific period in time," it will include many subobjectives in each of the participating functions of research, development, manufacturing, marketing, and administration. If an organization prefers multifunctional teams, subobjectives must be developed. Rather than think in terms of functional projects, organizations must reorient their thinking to product projects that enhance business performance.

3. *A well-integrated project plan.* The plan is not the flow chart that shows the critical path. A plan in that sense involves the thinking that must go into determining how the many different functions will interact in the related activities. Who needs what information from whom and at what time? How can the deadtime in waiting for resources and information be eliminated? The objective here is to draw attention to the critical issues that affect performance.

4. *A delineation of the critical success factors in each of the participating disciplines and functions.* Search for the issues that must be resolved for the project to meet the expectations—the issues that, if not resolved, could lead to failure and a waste of limited resources. Solve the supposedly unsolvable first. Prioritize the critical issues by their importance to the project. As an example, identify if the unavailability of some component, raw material, or information could result in project failure; identify the levels of competence as suggested by Kandel, Remy, Stein, and Durand, in "Who's Who in Technology: Identifying Technological Competence within the Firm" [4].

5. *Focus on the up-front activities.* Think before doing, and then think and do. Adequate resources are essential, keeping in mind that there are ten. Although people are important, the other nine resources also must be available.

6. *Understand the role of each of the participating functions.* Everyone obviously understands them? Not so. If they did, manufacturing would not be solving product problems in the factory that should have been solved in research, design, and development. Role definition, subobjectives related to performance and timing, and so forth, determine level of success.

7. *Effective project managers emphasize value-adding activities.* Where is the time that is being charged to the project spent? What are the contributions from the time charged? How have those contributions enhanced the value of the total project? As an example, expediting and coordination are not value-adding activities; they have been paid for in the past.

Figure 6.2 Basic requirements for successful new product project.

resources, and so forth. Those conditions apply to projects related to technology, products, processes, marketing, and acquisitions, those in the administrative functions, and expanding into global markets. Good ideas, without workable or state-of-the-art technology and without adequate resources to convert them into viable products, consume time that could be allocated more appropriately to projects that add value. Good ideas that are not salable only divert resources from more advantageous activities.

Within the project cycle time, projects achieve various levels of success; some surpass the requirements and others never achieve the objectives.

To optimize project cycle time, projects must take on a more integrated approach. Every organization has its share of projects in the functional groups. They are usually managed independently in the hope that at some time in the future the independent results can be reintegrated into a working system. That approach has not been very productive. Increased project cycle time can be traced to the problems associated with managing the interfaces between the functional groups and upper-level managers and the inability of the latter to integrate the activities of the former.

Preapproval considerations

Optimizing project cycle time begins with the many activities that actually precede the approval process. Although the project management functions after approval are important, it is difficult to optimize the cycle time of a project that was approved without adequate information or the exploration of viable alternatives. It is not uncommon to hear such questions as how or why did we ever decide to go ahead with that project? Too often the original design concepts that were used to justify an investment are ignored after approval. That is true of both product and capital projects. Arbitrary changes are introduced without communicating the changes to interested participants.

Considering the project cycle time as continuing from concept to implementation is the managing technology approach [5]. This substitutes the systems approach for the more usual helter-skelter process that often ignores the strategic directions of the business. Optimizing project cycle time is not a matter of project management mechanics or methodologies. The tools for managing projects are numerous, and there are varieties to meet many different needs. Although application of those tools is important even in the preapproval process, the quality of the information used in making decisions becomes much more important. Understanding the soft data, which includes the qualitative, the undefinable, the speculative, and knowing how to extract the substance from the data, determines the probability of success. The data must be transformed into validated information before a *number* can be assigned. That is true of all marketing information as well as the cost and performance projections emanating from research, development, manufacturing, and marketing.

Optimizing project cycle time assumes the selection of appropriate projects. There are no mysteries associated with project selection, but neither are there any detailed prescriptions or expert systems that will provide answers. A typical prescription might include considering the strategic and business fit, the marketing effort, the technology evaluation, a resource capabilities study, the need for new manufacturing facilities, the requirements of the distribution system, some legal considerations, and a financial analysis. Such a prescription would be a straightforward analytical approach followed by the typical approaches to decision analysis and synthesis. The difficulty in making the appropriate decision stems from the fact that most organizations do not take the time to validate the data they transform into information. It is further complicated by the fact that rather than evaluate alternatives, managers generally decide what they want to do and then find a way to justify the investment. Too often, managers fail to do an incremental financial analysis to determine which parts of a project can be justified and which parts are marginal or totally unjustifiable. That is not to imply that only the parts that can be justified will be approved, but it does tell management where the benefits are being generated. In the preapproval stage, emotion often overcomes reason and the shortcuts eventually are translated into costly but avoidable mistakes. There is no doubt that investment involves uncertainty and risks, but assumptions used in arriving at decisions must be questioned and analyzed as to their integrity, reality, and importance relative to the specific project under consideration.

Gaynor, in "Selecting Projects" [6], raises questions that must be asked during the preapproval period by all of the participants. They have to do with:

- Project objectives and understanding those objectives
- Business, marketing, and technology strategies and those of the competitors
- Importance of the project to the future of the organization
- Alternative approaches
- Differentiation between a business project and a technology project
- Business, marketing, or technological alliances
- Opportunities for leveraging organizational resources—value added
- Importance of proprietary technologies and competitive technologies
- Total cost to the business, not just the R&D
- The full extent of the resources required for implementation

Integrated Approach to Managing Project Cycle Time

Project cycle time includes more than R&D. Integrating the activities of research, development, manufacturing, and marketing is absolutely essential for optimizing project cycle time. The talent and knowledge base of each of the functions is involved in successful projects that are completed on schedule, meet the estimated costs, and perform according to specifications. None can be eliminated. None of the participants can take a back seat or sit on the bench waiting to be called. All must be active participants in the process by providing their input and making their specific contributions in a timely manner.

Organizations meet their objectives through projects. The management of those projects to reach the predetermined objectives in an optimal time period becomes paramount. But interdepartmental and interdisciplinary conflict often results in something less than expected. Elmes and Wilemon, in "A Field Study of Intergroup Integration in Technology-Based Organizations" [7], report on the major causes of intergroup problems. The conflicts stem from the many interfaces that have been created because of the microsegmentation of the organization. As Lawrence and Lorsch remind us in *Organization and Environment* [8], better-performing organizations are highly differentiated and highly integrated. They are highly differentiated, and they assume different cultural characteristics and orientation to cope with their special needs. They are integrated in that, despite different orientations, they are capable of functioning as units.

Research

"Research," as used in the project cycle time context, means research that must be done in relation to some specific new-product introduction. It does not include activities that might be part of a central research laboratory of a large organization or exploring the unknown for some possible future development. In most cases the research could relate to specific materials or the adaptation of materials for specific purposes. As an example, it could include the modification of polymers but not the search for a new class of polymers that may take years of work to discover.

The research effort must also be differentiated from organizations involved in the process industries, those involved in the discrete parts manufacturing, and those solely involved in assembly of mechanical or electronic components. As a general rule the process and the process-related industries require more research effort than the discrete parts manufacturers. Photographic film and magnetic media are two examples of new-to-the-market products that require time-consuming research. It took Kodak about 10 years to change from the 126 format to the 110 format. The change included new emulsions, improved coating and drying facilities, a new line of cameras, new facilities for developing and printing, and all of the changes in the infrastructure necessary to sell the film and provide finished prints to the customer. A similar statement applies to magnetic media, whether computer tape and disks or audio and video cassettes. Improving magnetic media density requires continual major improvements in the manufacturing processes. The changes begin in the laboratory; they are not just introduced on the factory floor.

Design and development

The development process involves providing a workable model that can meet the agreed-upon specifications. It includes testing and evaluation either internally or at some designated user's facility to determine the reliability of the product. If related to a process-type product, it must be demonstrated in a pilot plant that provides the information necessary for scaling up to the production facility. Too often, products are scaled up before the major problems have been eliminated. Even though this is the end of the twentieth century, organizations continue to think that defects in design can be resolved on the factory floor. Managers must recognize that each generation must be educated.

The development function includes design; generally, design is given second-rate status to development. What is generally classed as development includes design. Design requires creativity; it involves use of the imagination and the best mental processes available. Ideas are plentiful, but applicable and usable ideas are difficult to find. Much of the effort that goes into developing an idea is really design at its very best, and not handbook engineering. Someone must transform feasibility or laboratory mock-ups into products that can be manufactured economically at acceptable quality levels.

Manufacturing

Research has shown that too often manufacturing is not considered during the preapproval activities, with the result that management must respond to many surprises that could have been anticipated. The surprises often lead to costly investments of capital equipment that may not only lengthen the project cycle but significantly reduce the return on the total investment. Cost of manufacturing facilities must be considered in the project planning process.

The statement has been made previously that manufacturing begins in the laboratory. That statement must be reiterated. It is too late to begin thinking about the manufacturing process when the product has cleared the development hurdle. That is true of the process industries and the discrete parts or component manufacturers. If manufacturing is not brought into the process early enough, both research and development may spend excessive time in their activities. As an example, in one specific situation a major development effort was extended by almost 2 years because the development group thought the product would have to be produced with the current manufacturing equipment. If manufacturing had been involved in the early stages of the project, some relatively minor changes in the manufacturing process could have been made and the development effort could have been reduced by 18 months.

Marketing

The role of marketing in the new product development process continues to be debated. Marketing is not just the market research arm or the market survey operation. It includes sales and the many other activities related to determining the products to be marketed such as pricing, inventory control, advertising, product introduction, physical distribution, and customer service. There should be no doubt nor should any proof be required that the marketing-related activities must be implemented concurrently with technological activities. Engineers and scientists are not the sole arbiters of what functions should be introduced to a product. Technological elegance does not necessarily produce marketable products. Some bells and whistles may be needed, but generally not many. The difficulty that occurs too often is that marketers do not understand the full implications or the expected contributions of marketing, nor do they understand the related technologies. Although marketers need not be specialists in technology, they should understand the technologies involved in their products and what those technologies do for the customer. An organization may sell products and not the technology of those products, but understanding the limitations that the technologies impose on the ultimate use of the product is an essential marketing tool for the salesperson.

Strategic and Operational Issues

Managing project cycle time is neither a strategic nor an operational issue; it is both strategic and operational. Project cycle time determines the effectiveness and the eventual profitability of new-product introductions. Adler, Riggs, and Wheelwright, in "Product Development Know-How: Trading Tactics for Strategy" [9], argue for the linkage of technology management and product development, but product development is accomplished through the project approach. Their study of 24 companies identified five key elements common to technology management and product (project) development:

- Assignment of management responsibilities
- Role of the functional departments

- Nature of planning
- Means of protecting the competitive advantage
- Linkage between technology management and product development

The first four elements are common to all activities relative to developing a viable business strategy and planning the allied operational activities. The fifth, the linkage between technology management and product development, is not a surprise conclusion. Not only are technology management and product development different sides of the same coin, their activities are continually intermingled in such a way that the two become the coin and give up their individual identities. There is little reason to manage technology in a product-driven company unless that technology relates to the organization's product base. Likewise, new-product introductions cannot be sustained without managing the related product technologies. The companies that Adler et al. studied in regard to the five elements listed above ranged from the tactical and compartmentalized to the strategic and integrated. Their *"tactical"* (operational) and my preference, "operational," are the same. They concluded that the tactical hinders the maintenance of a sustainable competitive advantage and the strategic approach positions the organization to take advantage of the technology's potential.

In considering the linkage between technology management and product development, they conclude that the strategic perspective sees technology management as a learning process that requires an iterative and interactive exchange with product development. They argue that the tactical approach to technology management isolates technology from the day-to-day operations and the new-product activities.

Personal experience and the experience of colleagues is that effective product development requires optimizing project cycle time. The arguments whether managing project cycle time is a strategic or operational (tactical) issue is academic. In the real world, in which schedules, costs, performance, and timing determine success, project cycle time is both strategic and operational. If strategy is described as the means by which an organization goes from where it is to where it wants to be, managing project cycle time will either help or hinder the process. Selection of products is more closely allied to business strategy, and successful implementation depends on operational practice and discipline. If strategy is to make an impact on business performance, it must be accompanied by operations that provide and utilize all the resources available to the organization.

Linkages of Technology, Marketing, New Products, and Cycle Time

Adler, Riggs, and Wheelwright [9] link technology and product development. Marketing must be added to that duo. In reality, the total organizational effort must be devoted to new-product development activities. The marketing function, because of its close contact with the customer, has opportunity to provide the direction of product development, which can greatly affect the future of the organization. What is occurring in the marketplace is not a topic for the monthly report: It is information that should be communicated immediately if it has any bearing on current products or products in the development pipeline. To fulfill their role, marketing people must keep their antennas tuned constantly to the changing needs of the marketplace not only for improvements in current products but also for new products. Marketing activities require people who are creative and innovative if they are to benefit the organization. The sales organization is not just an order-taking department. If its participants are educated to recognize useful information, they become a valuable information resource for the organization. Creative marketing goes far beyond developing the best multicolored sales literature. The creative element makes the difference. The remainder is routine.

Linking technology, marketing, and new products involves taking a holistic approach to managing the trio as one. All organizations have their share of dedicated individuals who are willing to go the extra mile. The objective is to get them to play the same tune. What is needed is not a product development or multifunctional team, but an *organizationally driven team* (project team) as suggested by Walsh, in "Get the Whole Organization behind New Product Development" [10]. Walsh suggests that an organizationally driven team is built around four concepts:

- A holistic philosophy of new product development. That means that all the functional groups are appropriately involved in the up-front work and not after project approval.

- Date making and date keeping. In other words, do what you agreed to do with no excuses. That requires some forward thinking and anticipation of potential problems.

- Individualized commitment based on education and training. Education means that executives and managers have detailed understanding of the technologies involved in their products and functional groups have an understanding of the business objectives, product costs, timing, market targets and so on.

- Project control by means of a *new-product integration manager.*

Walsh suggests this manager is essential to coordinate the activities of the functional groups and should be at a level comparable to that of the functional managers, but coming from one of the staff groups.

Organizationally driven activities, whether related to new products, processes, services, or building new factories, generally prove to be successful. Every successful organization, at one time or another, has looked back with pride on some major accomplishment, and not realized what factors were responsible for its success. The organizationally driven approach focuses the activities not just of research, development, manufacturing, or marketing but of the total organization including all the administrative functions shown in Fig. 2.2, the tripartite organization. Although the concept of organizationally driven teams may not be new, Walsh's approach should cause managers to reflect on the circumstances that were responsible for their past successes and failures and consider the potential benefits of engaging the whole organization in the process of optimizing project cycle time.

Key Issues in Optimizing Project Cycle Time

There is no lack of books about project management. Most relate to project planning, scheduling, and control. There is no doubt that those are essential elements of the process, but by themselves they are insufficient. It is more important to understand what is to be planned and how the resources can be optimized and scheduled. Financial and scheduling controls do not guarantee results; they are history. Real-time control processes, or, more appropriately, real-time knowledge and understanding, involve first-hand knowledge of what is being accomplished, where the current effort is directed, and potential problem areas. It requires anticipating and solving potential problems before they occur. Control, to be useful and value adding, cannot be performed on a monthly or quarterly time scale. If a group of engineers spends a month going in the wrong direction, it can spend several months getting back on track. Some key issues in optimizing project cycle time must be emphasized:

- Project objectives and specifications
- Consistent priorities
- Agreement on timing
- Project resources
- Project organization
- Communication

Those issues may seem obvious and trite, yet they continue to be ignored. Furthermore, each is interpreted differently and usually without sufficient thought about the activities, resources, and infrastructure of the organization.

Project objectives and specifications

Developing objectives is not a new concept; on the contrary, it has been around for millennia. It may seem elementary and trite, but too often project objectives are not developed and communicated to the organization. If they are, they are given in general terms without sufficient forethought. As a simple example, consider the situation in which the manager describes the objective as building a fence. What kind of fence? For what purpose? To protect whom against what? At what cost? With what types of material? Temporary or permanent? Setting such an objective as building a fence could surprise the individual who established the objective. The type of fence, the cost, the timing, and so on could take on many different combinations. Managers continue to establish objectives without sufficient understanding of the implications of those objectives on the system. Most managers claim to be objective-oriented, but personal research has shown the opposite. The illusion must be replaced by the reality. Most of the time, objectives lack specificity, include divergent subobjectives that are not identified, are not clearly delineated, and lack the yardstick against which performance will be measured.

Project objectives and product specifications are considered jointly because they are codependent. Objectives basically ask questions: Why are we doing this? Why are resources being allocated to this? Specifications define or describe the *thing to be done.* What characteristics will this thing exhibit after it has been developed? Lack of clearly delineated specifications causes a great deal of grief, misunderstanding, added cost, and extension of project cycle time. Much of the current literature emphasizes the cost of engineering changes, but it is not the cost of the engineering changes that is important, it is all the added costs that the engineering changes create. Relatively speaking, engineering time costs little, but once the component or assembly is released to the factory, the cost increases rapidly. It is no longer a matter of engineering cost alone; it is the costs associated with many people throughout the organization and the cost of raw materials, energy, and production facilities. Effective project management, which ultimately determines project cycle time, relies on product specifications. There is no doubt that, on long-term development projects, specifications might require changes, but those changes must be anticipated by the participants, agreed upon, and communicated to all participants—including the customers who were contacted for input on the product specification.

Consistent priorities

Project priorities must be consistent with organizational priorities. Walsh [10] raised the issue of organizationally driven product development. Project cycle time can be optimized only if all the participants meet their commitments. The reality of everyday life mitigates against meeting that objective. Usually, organizations are not dealing with a single project. That would be the rare exception even in a small company. As an example, how does a function such as purchasing meet the different project needs in a timely manner? Every project manager considers his or her project as the number one priority. Corporate management must define the priority projects and plan accordingly. Completion dates cannot be established and the project participants cannot be held responsible for performance if the supporting functions give it a low priority.

In reviewing an organization's work load schedule, the analysis showed that some projects had been on the project list without any activity for 3 or 4 years. Priorities must be established, but the simplest technique requires determining only whether the project is active or inactive. If it is active, it is a top priority project. No number 1, 2, 3, 4, or 5 priorities. Companies that use a numerical approach for establishing priorities seldom work beyond numbers 1 and 2. The excess paperwork is carried from month to month without adding any value. There are no available resources to pursue every activity regardless of its desirability. Choices must be made; that is the role of all decision makers regardless of level in the organization.

Agreement on timing

Agreement on timing involves establishing business priorities and understanding the availability of the business resources. There is no doubt that the marketplace establishes the timing for introducing a new product that determines the project cycle time. Entering the market too early can be just as detrimental as entering it too late. However, the marketplace timing often requires an organization to change priorities in order to maintain its market position. Such situations, if they occur frequently, can undermine the confidence of the participants in their management. Some education is required on both sides of this fence. Those involved in projects must recognize that priorities will be changed, and management must look far enough into the future to minimize any unnecessary disruptions in the work routine.

The other aspect of timing concerns the timing of specific activities within a project in order to reach the stated objectives. A director of a product development function once stated that he does not believe in scheduling of any type, even the use of simple Gantt charts. Such an attitude cannot be supported by the typical profit-making organization. The attitude only reinforces the concept of functional segmentation and the development of fiefdoms that place greater emphasis on their own performances than the performance of the complete organization. Functional groups interact, and the information generated by one group usually must be passed on in a timely manner to another. There comes a time when tooling must be designed, but that tooling cannot be finalized without the design information. Ignoring the information timing needs clearly mitigates against optimizing cycle time.

Project resources

Part of the difficulty lies in the fact that most organizations have too many active projects and lack the resources to meet their objectives in a timely manner. This situation results in inadequate staffing, not just in the specific project but in the support functions. Balancing the number of projects with the available resources is a prerequisite to managing project cycle time. Assumptions about the availability of resources cannot be made. The resources are either present to the extent required or not. As a matter of reference, those resources include people, intellectual effort, information, culture, customers, time, technology, finances, plant and equipment, and facilities.

The limitations of those resources must be recognized. The knowledge of each must be factual, and the discrepancies between the needs and the availability must be understood. The idea that somehow we will manage to get the resources we need does not result in meeting the project objectives. Usually, objectives are met by taking resources from other projects that may be equally important. Once the cycle begins, using resources from other projects results in a never-ending stream of reallocations. Such activities merely build to the point at which project schedules become completely meaningless. Time is spent juggling and fighting for resources rather than optimizing their use.

Very often, expenditures are approved too far in advance of the resources. As an example, an executive was boasting about the number of dollars of approved capital investments. Some projects extended beyond 4 years. When asked how he planned to cope with the added work load, he shrugged his shoulders and said that in due time all the projects would be completed. That situation presents management with some real challenges. Some questions must be asked about the number of projects that can be sustained by a work force in order to maintain a constant effort to complete what has been started. How many projects have not reached their performance objectives? How many changes were never recorded in the documentation? How many projects are behind schedule? What is the net impact on the future of the business if the project is delayed? What about just approving the engineering effort? If the resources are not available either internally or externally or if the financial resources are limited, why approve such projects? Logic as well as financial considerations would dictate approving only investments that can be realized with the available resources and within a timely manner, whatever that time requirement may be.

Project organization

There is little doubt that project organization plays a role in optimizing project cycle time. The physical organization is not the important issue. Project organization includes the attitude with which the participants approach their activities and their relationships. Whether the groups of specific talent are called task forces, teams, cross-functional teams, or multidisciplinary teams or are assembled in a matrix-type organization does not make a great deal of difference. At least there is no quantified research to demonstrate that any specific project organizational structure will be more effective than another. People are not a constant. They act, react, and interact differently and often unpredictably. What is important from an organizational viewpoint is how the participants perceive their roles in the structure.

The approach by Walsh [10] of organizationally driven product development has nothing to do with boxes on a chart showing levels in the project hierarchy. In an organizationally driven project, levels and functions lose their importance because the pursuit of the end result keeps the group focused on the objective. That does not mean the physical structure should not be considered. What it does say is that the attitude with which the participants approach their activities is far more important than the boxes on the formal project organization chart. There are no anointed princes in an organizationally driven project.

Within an organization, managing a single project from an organizationally driven perspective may or may not provide an advantage. Organizations are usually involved in many projects. It may be relatively easy to single out one project as organizationally driven, but it must be recognized that all projects regardless of size or technological scope, must be managed as organizationally driven. That must be the target for organizations that hope to maintain a competitive position; it is a target for optimizing project cycle time. Furthermore, there is no better way to motivate an organization than through success. Businesses achieve success by effective management of many projects simultaneously in all the functions. Success begets success until it becomes arrogant and hubris begins to destroy the inner motivation that was responsible for it.

Walsh identified the need for a new-product integration manager to coordinate the functional activities. However, such an appointee can only lead to future problems. Appointing people to coordinate what others are being paid to do violates one of the cardinal principles of effective and efficient management. Coordinators without power to act only add another level of unproductive management. They usually do not understand the technologies or the markets. *The coordinators who are responsible and accountable for performance and have the authority to make decisions are the project managers.* They are in a position to put it all together.

The current lack of performance in optimizing cycle time can be traced to the fact that organizations have not yet recognized the importance of the role of the project manager. Competent project managers possess some very definite skills. Project management deserves the status of a profession. In many situations, a project manager requires multidisciplinary education and experience to function effectively. Project management is a profession that requires very specific skills and talent, both in depth and in breadth. People who can understand the needs of research, development, manufacturing, marketing, and the business are few and far between. Those who can link the functional disciplines with the tools for project management are even scarcer. In the final analysis, the project managers determine project success or failure.

Project managers, keeping in mind that they span the spectrum from leading edge scientific research to implementation of often mundane and routine objectives, assume the characteristics of what are often referred to as product champions. Howell and Higgins, in "Champions of Change" [11], describe champions as the "individuals who emerge to take creative ideas, which they may or may not have generated, and bring the ideas to life." Their study included 28 successful innovations in information technology. It related leadership behavior, career experiences, and the three processes by which champions operate: the renegade, the rational, and the participative. Although their research is limited to champions in information systems, my observations over many years in technology-related management have convinced me that the findings extend to champions in all areas and disciplines. Those they refer to as champions also include project managers. Champions and project managers exhibit similar characteristics. A champion must be a project manager, and a project manager must be a champion in the truest sense. Both are presented with a continuous barrage of obstacles.

It is important to recognize that the success factors may change during various time periods in the project life cycle. Pinto and Prescott, in

"Changes In Critical Success Factors over the Life of a Project" [12], concluded after a study of ten success factors that "the relative importance of various critical success factors are subject to change at different phases of the implementation process." The findings imply that success factor analysis may be contingent upon project life cycle. They also noted that "the assigned project manager may be in a better position to assist in the implementation, given the increased awareness of the factors most critical to success at specific life cycle stages."

McDonough, in "An Investigation of the Relationship between Project Performance and Characteristics of Project Leaders" [13], supports the hypothesis that project performance is affected by cognitive style, the career orientation, and the background characteristics of the project manager. The performance is also affected by the type of work undertaken by the project team. Cognitive style relates to the manner in which an individual acquires, filters, and analyzes data and then synthesizes the information prior to making a decision. Creative ability and innovativeness have been shown to be associated with project quality, but the level depends on the type of project. McDonough's research clearly indicates that different types of project leadership are needed for different projects. There is no "one best," and the type that is most effective depends on the work to be undertaken.

My informal investigation over many years and in many organizations indicates that the role of the project manager is temporary (for the duration of the project). On completion of a project, instead of being assigned to another project, the manager returns to some other activity. That is true whether the project relates to products, manufacturing facilities, or one of the other types of projects in the business. Starting every project with a new project manager is a costly endeavor and a misuse of business resources. Everyone starts out at the bottom of the learning curve and fumbles through the same series of mistakes.

Communication

Inadequate communication is not the sole source of extended project cycle time, but it is an important consideration. Communication threads its way through the other key issues; it is really information transfer. The familiar and often heard statements of "tossing it over the wall" or "passing the baton" are forms of communication, although they are totally ineffective. Good communication is immediate and timely; it conveys the information essential for meeting the performance requirements. Knowledge of the purposes, objectives, and strategies of the organization is communication and information transfer. Educating the production people about the customer complaints on a regular basis is communication of important information. Mutual interchange of the required information between the functions involved in a project in *real time* is communication.

Organizations tend to complicate the information transfer process. Generally, there is too much emphasis on so-called confidential information. The guidelines for what should be communicated or what information should be transferred is very simple: It should be based on the *need to know*. However, the need to know must include everything that individuals need to perform their jobs efficiently. It requires a broad interpretation rather than hair splitting by management. Too often, the so-called confidential list eliminates many of the individuals who really need the information. As an example, although management often prefers to consider costs as confidential information, by not informing employees of the costs related to or associated with their activities, it severely limits the opportunities for improvement. If the desired cost of a design is not communicated at the beginning of the design process, cost will be added through redesign or engineering changes that come about because of a lack of that cost information. As an example, the acceptable and approximate manufacturing cost (a target figure) of a product should be known at the time of design and development. Usually there are alternative designs that can meet the cost targets if those cost targets are known. If design and development are not provided with these target figures, elegant designs with the latest technologies may take precedence over providing design integrity with adequate technologies, that is, technologies that are proven and meet the requirements. Redesign effort extends project cycle time.

Lessons from a Case Study

Project cycle time embodies a large number of considerations. Cohen and Knospe, in "Professional Excellence Committee Benefits Technical Professional at DuPont" [14], report on on how DuPont technical people in one division credit a committee with improving risk taking, teamwork, career self-management, and communication. The committee was organized with the aim of providing a forum by which the technical professionals could determine how they could contribute more effectively to the overall goals of their departments. At their initial meeting, they identified the major issues and concerns of the group and rated them on a scale of 1 to 10, zero representing the not important and 10 the important. The results of the study are shown in Table 6.1.

This committee activity represents a good first effort. It may at least have a positive impact on research projects. As was stated, the purpose was to determine what could be done to affect the *overall goals of the department*. It is necessary, however, to take the next step and deter-

TABLE 6.1   Findings of the Professional Excellence Committee at DuPont*

Major issues and concerns	Rating
Lack of effective teamwork between research and development, marketing and manufacturing	8.6
Ineffective inter- and intragroup communication	8.4
Lack and fear of risk taking	8.2
Lack of effective career planning	7.9
Lack of understanding the marketplace	7.8
Ineffective technical assignment	7.7
Lack of forums for proactive thinking	7.5
Too much firefighting of plant problems	7.4
Insufficient research time because of administrative duties	7.2
Individuals unaware of business goals	7.1
Too much inertia and resistance to change	6.9
Some individuals not self-motivated	6.8
Too much tolerance of poor performers	6.7
Insufficient recognition	6.6
Priorities not clear or change too often	6.5

*The committee was organized for the purpose of determining how technical people could contribute more effectively to the goals of their departments.

mine specific changes the research department must make to *contribute more effectively to the overall goals of the corporation.* Meeting department goals is insufficient; they do not necessarily optimize the project cycle time for introducing a product to the market. The possibility is to have a research organization that is effective from the departmental perspective but is not contributing to new product introductions as it should. Such a committee effort looks inward, but it must also look beyond its defined responsibilities.

The 15 major issues identified in the DuPont study would be the same in most technology-oriented organizations: they impact project time, project duration, and timing. Lessons can be learned from reviewing the factors that contribute to a lack of effectiveness and inefficiency. The extent to which each of them can be verified is exceedingly important. Just what is meant by such words or terms as "lack of," "ineffective," "too much," "insufficient," "unaware," and "lack of clarity" must be stated specifically. "Lack of" to one person can be "too much" to another. Such studies must eventually verify the comments for any meaningful change to be implemented.

Although the study rates the importance of the major concerns, it is important to recognize that none can be ignored. Single-issue management is not the answer. Lack of effective teamwork of research, development, marketing, and manufacturing is considered the most important, but the issues related to system inertia, tolerance for poor performers, and inconsistent and changing priorities must be resolved. Good teamwork depends upon the resolution.

The DuPont effort has been presented because many organizations undergo similar processes not only with their R&D departments but also with the other functional departments. The result, although providing some advantages, obscures the fact that R&D, like all other functions, are part of a business, not independent functions. Expanding the DuPont approach to each function and then synthesizing that information for the business unit into a system for managing provides an approach for optimizing project cycle time.

Managing the Enterprise with the Project Approach

The project approach to managing engineering and its related activities provides much of the discipline that is required to meet the requirements of performance, scheduling, and cost. It is used extensively in product development and to a far lesser degree in research and manufacturing. It may be used in marketing occasionally but seldom in the administrative functions and upper management levels. But what are the potential advantages of introducing the project approach throughout the organization, beginning with the CEO? Does that sound absurd? Is it a means of control, or can it instill the mental and operational discipline that is lacking in today's organizations? If it works well on the technical side of the business, why not adopt it throughout the organization?

Organizations tend to operate from crisis to crisis. The complaints about overwork in the halls of industry grow louder, but nothing significant is being done to change the system that supposedly is creating the overload. This may be a good time to begin looking at the potential for using the project approach. First, it is necessary to note that most organizations create their own crises and often enjoy the eternal state of confusion and uncertainty. It underscores a sense of mission and keeps everyone guessing.

Managing the activities of an organization is at the very least a complex and demanding activity. An organization does not resemble a machine that has specific functions, carries out deliberately designed actions controlled by computers, and produces components or assemblies to specifications. Organizations must be designed, but, unlike the product approach, the designs continue to change because of the dynamic environment in which they operate. Organizations emerge. They evolve on the basis of the political, economic, and social environment in which they operate. A hypothesis is offered for consideration: *Introducing the project approach throughout the organization is positively related to performance.* That hypothesis is an extension of the results attained with effective project management in design, development, and construction. If the technology-related functions can ask: "How can we reduce the cycle time by 50 percent?," can the same question be asked of the total organization?

The underlying principles for introducing the project approach throughout the organization for optimizing the three elements of cycle time (time, timing, and duration) requires responses to what, why, how, who, when, and where. The answers are very simple:

What	The project approach takes the knowledge gained from project management in design and development and unfolds it throughout the organization.
Why	Asking why introduces a mental and an operational discipline that directs the attention of all the participants to the importance of time, timing, and cycle time—earnings on the investment in resources.
How	Implementation is a learning process. It is not this year's major emphasis program; it requires altering the current way of thinking about doing things.
Who	Every employee becomes a participant in the process; there are no exceptions.
When	The project approach can be implemented when the infrastructure is capable of supporting it—step by step with an occasional giant leap as the organization climbs the learning curve.
Where	In every business-related operation whether internal or external to the organization; it includes suppliers and all outside business activities

The project approach: Why and why not?

Why should this project approach be used for all business activities? It provides a discipline and a systematic way, without strangling the participants with procedures, to focus attention on the important issues that include:

- An *objective* considered worthy of investing business resources
- Some understanding of the *level of investment* and the associated benefits
- A *process* for achieving the objective
- The short- and long-term *consequences* of the investment
- A means of *measuring* performance against the original objectives

These five conditions can be designated as objectives, investment, process, consequences, and measurement (OIPCM). If every nontechnology project in an organization were measured against OIPCM, staff groups could be reduced and many hours could be saved. The major concern is not the hours spent by the staff groups, but the hours spent in the profit centers as a result of staff group actions. That is not to say that staff groups are not essential, but, like any other investment, their activities must be justified.

It is interesting to speculate how organizations would have approached the investments in strategic planning, portfolio management, performance management, total quality control, just-in-time manufacturing, time-to-market cycle time, simultaneous engineering, managing technology, the Baldridge Award, leveraged buyouts, acquisitions, and many other programs if the project approach had been in force throughout the organization. The results would have been far more satisfactory. Some programs may never have been approved. Businesses are not charities. If the project cannot be justified, it has no reason to survive, regardless of its origin. That justification may involve different degrees of logic, rationality, and emotion, but it must be evident. It cannot appear to the organization as a roll of the dice.

Justification of a project can be based on either quantitative or qualitative measures. As noted previously, assigning numbers to qualitative information does not necessarily make that information quantitative. Quantitative information is a documentation of past history; it is a measure of past performance. The future is speculative and uncertain, and although numbers can be used to project future performance, they must be viewed as qualitative: They represent an extrapolation of the past plus some added insight to the happenings of the past and how those events will govern the future. But qualitative information need not be suspect if it is substantiated by the best available information. It is more difficult to work with, but it provides greater understanding. As an example, it may not be possible to put numbers to the benefits achieved by implementing a new personnel appraisal system, but the benefits must be identified. That benefit must be described in terms that go far beyond the statement, "We think it's a better or more equitable approach." Questions must be asked and answered. What does it do for the company? What does it do for the employees? How does it allow the organization to meet its purposes and objectives according to the three E's? Just what is this new approach going to accomplish? The list must be specific, and the benefits must be described in concrete terms—not better career guidance, but the ways in which the new program will improve career guidance, what improved career guidance is going to do for the individual and the organization, and how the results will be measured. The introduction of that new appraisal system will affect cycle time.

A response to "why not use the project approach throughout the business" is equally important: What are the major negative consequences if it is not implemented? Businesses do not operate anywhere near their potential, primarily because of a lack of mental and business discipline. The project approach imposes a discipline. It basically says: "Focus on the priorities and on the objectives." It provides a path for integrating the individual fiefdoms back into the kingdom. It is a means of integrating all the associated functions into a holistic organization. At the same time, it sends a strong message to the organization: This is one organization with specific purposes, objectives, and strategies and not a menagerie of independent functions each with its own agenda.

Implementing the project approach within the whole organization should not expose the organization to any new programs. Executives need only look at how their organizations currently approach technology, product, or other projects. In the broadest sense, implementing *anything* involves prior knowledge and learning in the process of implementation. If "knowledge is expertise," as described by Cleveland [15], then it is important to know where that expertise exists within the organization. If it is not available, it must be acquired—preferably by recruiting a person who understands the process and the problems associated with the general topic of project management. The project approach could begin as a pilot project, or it could be rolled out to the whole organization. It depends on the size of the organization and the understanding, support, and confidence that the potential benefits will become a reality. Since the process requires a behavioral change, there must be some evidence that the system will benefit the participants and the organization. Behavioral patterns are not changed by edict; each pattern change requires its own implementation plan. There is nothing better than knowing that the thing that management wants to implement is working effectively in other organizations. Some verification of the approach may also come from outside sources, from other organizations that are using the process even though only in a pilot stage.

The project approach should not be considered if the proponents intend to rely on high levels of paper shuffling as the measurement of success. Management by objectives (MBO) was destroyed in many organizations because it became a paper mill. The approach does not involve tracking time to the nearest minute, nor does it involve reporting the information to top-level staff executives. The purpose is to let individuals and their managers acknowledge whether the time resource is being allocated appropriately: Which activities are not adding value? It does not involve microtracking each activity. The objective of the project approach must be kept in mind. It is not an accounting innovation. Engineers generally charge their time to projects. Lawyers and consultants also track their time in order to bill their clients.

Lessons from the past

Two examples illustrate the benefits that can be achieved when organizations choose to manage holistically and from a systems perspective. The first is General Electric's Electrical Distribution and Controls Division, and the second is the resolution of operational difficulties between the capital engineering and the manufacturing functions of a multidivisional organization.

General Electric. Lener [16] reports on the appointment of William Sheeran as the productivity czar at General Electric Co. Sheeran was formerly the general manager at GE's Electrical Distribution and Controls Division (ED&C). He is currently vice president of corporate engineering, production, and sourcing. In his new position, he is expected to duplicate what he accomplished in ED&C: annual productivity gains of at least 5 to 6 percent. With his staff at ED&C, Sheeran [17] reduced:

- The number of plant locations from six to one
- Customer delivery cycle from 2 weeks to 3 days
- Average work hours to produce a panel from 4.5 to 1.7
- Total cost by 30 percent
- Direct labor by 55 percent
- Customized unique parts from 28,000 to 1300
- Inventory from 45 days to 9
- The project load from 140 projects to 14 programs

ED&C's accomplishments demonstrate the viability of the project as well as the systems approach. Changing ED&C and making it into an effective and competitive enterprise is really one project with many subprojects. It focused attention on the objective and all activities were reinforcing rather than in opposition. This project was a business project, a product project, a process project, an information project, and an automation project. All the elements of the business were included.

The GE project provides some insight into how change was approached at ED&C: changing how people think and work. When Sheeran arrived at ED&C, the first thing he did was force design and manufacturing engineers to sit side by side and talk to one another about product development. Historically, there was a great deal of conflict between the two groups. The objective was to move the manufacturing considerations up front into the design process. Was everybody willing? No. People who were barriers to the process were removed. The most intractable, according to Sheeran, were the drafting people. Some did not accept computer-aided design. The solution was simple: remove the drafting tables. Drafting people no longer were needed; Sheeran wanted only designers. Those who were qualified came back; others, according to a *Business Month* article, found other jobs within GE or in the Hartford area.

Sheeran is obviously the sparkplug to put the change process into effect. He makes one comment that should be considered by all organizations that attempt to introduce a change in the way that an organization thinks and operates. Although he was appointed the czar at GE, he states that his long-term goal is to hasten the obsolescence of his office: "The ultimate would be that within five years the individual businesses have so developed their missions that we have no reason to exist."

That scenario raises an issue that is seldom addressed as organizations attempt to create change or make significant improvements across the board: "People who were barriers to the process were removed." Some personal research into why organizations fail to implement programs successfully supports Sheeran's comments regarding the removal of the obstacles. Too often, managers, either through fear of being criticized, insufficient backup from performance appraisals, or just inability to deal effectively with controversial issues, rationalize their inaction in removing people who are nonresponsive to the needs of the organization.

Sheeran's approach for creating change is simple. First, it is necessary to convert the top executives and then the middle managers who will be responsible for making the system work. As Sheeran notes: "We're creating disciples of change," but that does not occur overnight. Educating the middle managers includes not only lectures but sessions for applying the learning to the specific problems of the division.

Functional conflict. A chief engineer, on appointment to lead a capital engineering department, inherited over 400 projects of various sizes and scope. His predecessor had requested additional staff of 15 people to cope with the project load. The newly appointed chief engineer chose to cancel the request for additional personnel until he had an opportunity to review the project load and the capability of the staff. After a considerable amount of study, he found that many of the old projects were in direct opposition to the more recent requests. The new chief engineer then called in the various engineering managers and asked them to determine, with the help of manufacturing, which projects were no longer necessary.

The initial reply was that only 10 of the approximately 400 projects could be eliminated. Having had some prior experience in such matters, the chief engineer once again called in the engineering managers and asked them to reexamine the project load and, through a process of elimination and consolidation, reduce the total number by at least 75 percent. That was supposedly an impossible task, but he assured them that he would be willing to work with them and provide all the assistance possible. His advice to them was to cancel all projects over 2 years old with no activity regardless of the reason. That could make a good start. In the process, manufacturing raised the issues of the explicit necessity for completing certain projects that had been inactive for more than 3 years. The chief engineer told the engineering managers to ignore the requests from manufacturing temporarily. When they completed their study, they would meet with manufacturing for a full review, project by project. After about 2 weeks, the 400 projects were reduced to a total of 75 including some consolidations.

The general impression was that the meeting with manufacturing would conclude with accusations and finger pointing. Looking at the requests from manufacturing objectively often becomes clouded by past practices of engineering and manufacturing. To defuse the situation, the chief engineer suggested at the meeting with manufacturing that participants look at each project in relation to its value added and its absolute necessity for meeting product requirements and justify their requests. The objective was to separate the musts from the wants. The participants were in agreement, and the group was reminded that all engineering activities were recharged to manufacturing. That was not done with the intent of decreasing the engineering work load, but to make everyone understand that investments must be justified in some manner.

After many hours of discussion, debate, and argument related to justifying the investment of time and energy, manufacturing essentially approved the list of projects. Some were added and others were eliminated. That was the first time in years that engineering and manufacturing had met for a joint review of all projects. In the past, manufacturing requested and engineering accepted; no questions were asked. This meeting began to change the way in which engineering and manufacturing viewed their respective activities: They realized they were part of the same organization. It was the beginning of an alliance between manufacturing and engineering. There were some rough moments over the next 2 years, and individuals who became barriers to the process had to be removed. It was the beginning of an era in which major capital improvement projects would include inputs from research, development, and manufacturing from day 1 and participate in the process through scale-up. Engineers did not stop being responsible when the equipment was installed; they remained on the project as needed until the performance targets were met.

Creating change

Success in creating change is largely determined by how rapidly the participants learn and how that new learning is applied. Argyris, in "Teaching Smart People How to Learn" [18], provides some insight into the problem. He states that: "Every company faces a learning dilemma: the smartest people find it hardest to learn." His 15-year in-depth study was related to consultants, but as a group the consultants represent the increasing role of highly educated professionals in industry. The study involves mostly MBAs from the top business schools. The transfer of the results to industry may be more meaningful than Argyris suggests. He uses the word "consultants" rather loosely. Newly minted MBAs are not consultants. They are hired contract workers with some special analytical capabilities: They possess some information that can be applied. That same situation exists in industry: Professionals in many different disciplines are capable of storing various types and levels of information and some knowledge gained from the prior application of that information.

Argyris claims that professionals avoid learning because they are threatened by the prospect of critically examining their roles and performance in the organization. That in turn leads to defensive reasoning and behavior and becomes a prime obstacle in creating change. To break the defensive reasoning posture, Argyris suggests that learning must be directed toward reasoning productively, and that involves linking the education to real-world business problems. Argyris emphasizes reasoning productively, but it must be recognized that at times it may even be mandatory to reason defensively. Argyris assumes that all parties are equally knowledgeable. As an example, some defensive reasoning may be required if a technology program is being presented to an executive committee that is illiterate about the technology or its benefits to the future of the business. Changing an organization requires large amounts of productive reasoning. It is usually easy to defend the past, especially if the past has been successful. U.S. industry is now reaping the results of defensive thinking. Questioning the reasoning processes of others is usually not met with any great enthusiasm, although it has cost U.S. industry billions of dollars. Recent accounts of major staff reductions at such organizations as IBM [19] and General Motors [20] clearly demonstrate what occurs when bureaucracy begins at the CEO and executive levels. A final quote from Argyris may set the stage for a more realistic approach to learning as contrasted to absorbing information: "To question someone else's reasoning is not a sign of mistrust, but a valuable opportunity for learning."

Implementing the project approach allows CEOs, executives, and managers to direct their attention to value-adding activities. They too need priorities in allocating their time. They must dedicate some of their time to improving project performance at all levels; they must spend more time on productivity in their organizations. Hise and McDaniel, in "American Competitiveness and the CEO: Who's Minding the Store?" [21], report on the responses from CEOs about the importance of the following eight business areas: customer relations, financial planning, production/manufacturing, new product planning and development, research and development, labor relations, personnel management, and market analysis. Responses were received from 236 CEOs by inquiry of the top 1000 U.S. firms. Financial planning was the top priority. Although new product planning and development, research and development, and production/manufacturing were considered highly essential for future growth, the CEO reports of involvement in those activities did not correlate with the apparent level of importance. The survey showed that only a small percent of CEOs spent *considerable time* in these technology-related activities:

- New product planning and development (14 percent)
- Research and development (8.4 percent)
- Production/manufacturing (4.6 percent)

It is difficult to understand why, if new product planning, R&D, and manufacturing are so essential for growth and maintaining high levels of business performance, CEOs pay so little attention not only to their specific activities but also their interdependence. The activities alone control more than 65 percent of the sales value of production.

Ross, in "Synergistics: A Strategy for Speed in the '90s" [22], suggests that:

[If] people respect the principles of your business, they can react quickly to change and still make good decisions; instead of fruitless or misdirected action, they can produce hard, fast results. Organized and empowered to contribute their full potential to their organization's goals, they create an irresistible force. They create the force of synergy.

Ross also states that the dynamics of synergy include: Simpler is faster. Freedom and opportunity are essential. Trust is a source of intelligent and creative decision-making. Those statements apply to projects also. In reality, a business is only an accumulation of various projects. If organizations choose to optimize the three elements of cycle time, the managers must learn how to make the total greater than the sum of the parts.

There is little doubt that simpler is faster, but only after the simplification process has been completed. It is the up-front work that allows simplification to take place. Freedom and opportunity to act rather than react are essential ingredients for optimizing cycle time, but they must include competence at the appropriate level. Trust as a source of intelligent and creative decision making becomes a reality only when people are afforded the opportunity to learn in the process of making mistakes. Teamwork, empowerment, innovation, and risk taking are essential.

Projects involve managing groups of people. Teams, task forces, and committees, or people brought together in some structure from different functional groups, provide organizations with the means for resolving management-related problems. Traditional group development models suggest that work effort is accomplished in some orderly series of stages. That may or may not be true depending upon the circumstances. Practitioners know that such models are at best guides to project management. They are not solutions. Managing routine, intellectually nondemanding, and noncreative activities may fit these models. Once creativity and innovation and interdisciplinary knowledge enter the scene, those models lose their significance. Gersick, in "Time and Transition in Work Teams: Toward a New Model of Work Development" [23], reports on the unexpected finding that teams progressed in a pattern of "punctuated equilibrium" through alternating patterns of inertia and revolution in behaviors. The findings also suggested that progress depended more on awareness of time and deadlines than on completion of some absolute amount of work. Gersick recognizes the limitations of the study and cautions against blind acceptance of the hypothesis.

Although Gersick's conclusions differ from the traditional sequential and orderly models, it should be recognized that team composition, the scope of the project, the size of the project, the environment in which the team functions, and other factors determine the degree of orderly information transfer.

Summary

- A project involves any activity regardless of size or scope that can be described sufficiently to assign the resources for its accomplishment.

- Project cycle includes the time from authorization through complete implementation to fulfillment of specifications.

- Optimization of project cycle time requires the integration of all the related functions within an organization.

- Optimizing project cycle time is both a strategic and an operational issue. Strategy without an operational plan for accomplishing the strategy provides no benefit.

- Although technology and product development are closely linked, marketing must at all times provide its input: Markets change; new competitive products are introduced; and customer wants and needs change. This is a dynamic world.

- Teams, if they are going to play a role in optimizing project cycle time and thus positively affect business performance, must be organizationally rather than functionally driven.

- Project objectives and specifications, consistent priorities throughout the organization, setting completion dates for meeting objectives, providing the resources, organizing the project, and effectively communicating are absolutely essential.

The project approach to managing a business unit requires the same discipline as that associated with managing engineering projects. It provides opportunities to:

- Develop a business mental discipline

- Influence the direction of projects originating in various function because of an understanding of the interrelations

- Apply some of the processes normally associated with the technology functions within the total organization

- Eliminate the ad hoc management approach—different principles on different days

- Approve the management-related projects based on the same principles as those related to research, development, and so on

- Take a systems approach

- Think time in terms of the business rather than the functions

References

1. R. D. Archibald, "Implementing Business Strategies through Projects," In W. R. King and D. I. Cleland, *Strategic Planning and Management Handbook,* Van Nostrand Reinhold, New York, 1987, pp. 499–507.
2. G. J. Ritz, *Total Engineering Project Management,* McGraw-Hill, New York, 1990, p. 4.
3. G. H. Gaynor, *Achieving the Competitive Edge through Integrated Technology Management,* McGraw-Hill, New York, 1991, pp. 139–163.
4. N. Kandel, Jean-Pierre Remy, C. Stein, and T. Durand, "Who's Who in Technology: Identifying Technological Competence within the Firm," *R&D Management,* vol. 21, no. 3, 1991, pp. 215–228.
5. G. H. Gaynor, *Achieving the Competitive Edge through Integrated Technology Management,* McGraw-Hill, New York, 1991.
6. G. H. Gaynor, "Selecting Projects," *Research Technology Management,* July-August 1990, vol. 33, no. 2, pp. 43–45.
7. M. Elmes and D. Wilemon, "A Field Study of Intergroup Integration in Technology-

Based Organizations," *J. Eng. and Tech. Management,* vol. 7, nos. 3 and 4, 1991, pp. 229–250.

8. P. J. Lawrence and J. W. Lorsch, *Organization and Environment,* Harvard Business School Press, Boston, 1967.

9. P. S. Adler, H. E. Riggs, and S. C. Wheelwright, "Product Development Know-How: Trading Tactics for Strategy," *Sloan Management Review,* vol. 31, no. 1, 1989.

10. W. J. Walsh, "Get the Whole Organization behind New Product Development," *Research/Technology Management,* November-December 1990, pp.32–36.

11. J. Howell and C. Higgins, "Champions of Change," *Business Quarterly,* Spring 1990, pp. 31–36.

12. J. K. Pinto and J. E. Prescott. "Changes in Critical Success Factor Importance over the Life of a Project," *Academy of Manage. Proc.,* 1987, pp. 128–131.

13. E. F. McDonough III, "An Investigation of the Relationship between Project Performance and Characteristics of Project Leaders," *J. Eng. and Tech. Management,* 1990, vol. 6, nos. 3 and 4, pp. 237–260.

14. E. D. Cohen and R. K. Knospe, "Professional Excellence Committee Benefits Technical Professionals at DuPont," *Research/Technology Management,* July-August 1990, pp. 46–52.

15. H. Cleveland, *The Knowledge Executive,* Truman Talley Books/E. P. Dutton, New York, 1985, pp. 22–25.

16. J. Lener, "The Right Tool for the Job," *Business Month,* April 1990, pp.62–65.

17. W. Sheeran, T. Campbell, and J. Sutton, "The Application of State-of-the Market CIM to GE's Electrical Distribution and Control Business," *Professional Program Session Record No. 11,* IEEE, Electro 88, May 1988, pp. 3–5.

18. C. Argyris, "Teaching Smart People How to Learn," *Harvard Bus. Rev.,* May-June 1991, pp. 95–105.

19. P. B. Carroll, "IBM Announces Details of Plan to Break Its Business Into More Autonomous Pieces," *The Wall Street Journal,* Dec. 6, 1991, p. B4.

20. J. B. White, "GM's Problems Have Overtaken Stempel's Go-Slow Approach," *The Wall Street Journal,* Dec. 16, 1991, p. B1.

21. R. T. Hise and S. W. McDaniel, "American Competitiveness and the CEO—Who's Minding the Store?" *Sloan Manage. Rev.,* Winter 1988.

22. A. Ross, "Synergistics: A Strategy for Speed in the '90s," *Business Quarterly,* Spring 1990, pp. 70–74.

23. C. J. G. Gersick, "Time and Transition in Work Teams: Toward a New Model of Work Development," *Academy of Management J.,* 1988, vol. 31, no. 1, pp. 9–41.

7

Time to Decision

Time to decision plays an important role in optimizing system, product, and project cycle time. It draws attention to the length of time needed to arrive at a decision by either an individual or a group. Time to decision has been singled out for special attention, since it is seldom considered by management as a major source in extending cycle time. It has not received any significant attention from academic researchers. Practitioners involved in any business activity should clearly understand the implications and the consequences of delayed decisions. Delays cost money, waste resources, and demotivate. Slow time to decision does not exist in the executive suites only. It permeates the complete organization at all levels. It affects the performance of the organization whether it involves the most important or the most trivial. It applies to making major decisions relative to the business as well as the thousands of decisions made in research, development, manufacturing, marketing, and all the administrative functions. The so-called minor and seemingly insignificant delays often negatively affect business performance. Time to decision applies equally to the strategic and the operational decisions.

Chapter 7 does not attempt to present a comprehensive picture of the many elements involved in decision making; instead, it focuses attention on how delays in decision making impact the business. The purpose is to provide some background to stir the imagination of executives, managers, and professionals and provide some insight into the potential negative impact on business performance when the time to decision extends to where business resources are wasted. Some of those decisions have a major impact on total business performance; others may only affect the performance of an individual or a small group. But collectively, the direct and indirect cost attributed to extended times to decision represent more than the net profit of most organizations. There are solutions for improving the time to decision, but first it is necessary to understand just how delays affect performance. Chapter 7 considers:

- Scope of time to decision
- Background information
- Decision models
- Decision support systems
- Impediments to decision making
- Factors affecting time to decision

Scope of Time to Decision

Time to decision means just what it says: How long does it take to make the decision? The only uncertainty is when time starts and ends. It begins with some deliberate action that is taken toward making a decision and ends when that decision is made. According to Mintzberg [1], "The decision process encompasses all those steps taken from the time a stimulus for action is perceived until the time a commitment to action is made."

There is no simple sequential process for decision making. Decision making has been described in many different ways. Lindblom [2] described it as the "science of muddling through." Others have attempted to describe it as consisting of various phases such as identification, development, and selection of viable alternatives. The decision process related to important or highly controversial or sensitive issues is a combination of some structure, some muddling around, searching for information, doing an analysis, and subsequently synthesizing a decision. The results of that complete process may be written or may be in the mind of an individual. It may require many months or a year or more, or it may be made in a very short time. Excluding decision makers who shoot from the hip or make decisions because any decision is better than none, most decisions go through a process of some kind with different degrees of discipline.

Hickson et al. [3] report the results of the process time for what they describe as strategic level decisions. They found that it took, on average, over 12 months to reach a decision. I question the use of the word "strategic" because of the types of decision involved: They were of the operational type. According to the study, the time to decision varied from 2 months to 4 years. The study includes a broad spectrum of organizations: private enterprises, academia, municipalities, and government agencies.

Background Information

Although academic researchers have studied decision processes of executives and managers, very little insight has emerged. Such studies usually occur after the fact. They are not made in real time, nor can they enter the minds of the decision makers. Decisions span such a large continuum that what may be applicable in one situation may not be in another. People are involved in decision making, and every person brings different perspectives of the inputs to the process. Although the challenge may be to wrap decision processes up into a neat package, it is doubtful whether a generic approach could ever be achieved. Further difficulty stems from the integrity of the information about how decisions are made. A comprehensive study would be necessary; it would involve all the individuals who provided input or participated in the final decision, and even then the results might not be universally applicable. Dutton and Dukerich, in "Keeping an Eye on the Mirror: Image and Identity in Organizational Adaptation" [4], state that: "Image and identity guide and activate individuals' interpretation of an issue and motivations for actions on it, and those interpretations and motivations affect patterns of organizational action over time." Their work relates to organizational adaptation, but it provides an insight into how image and identity affect time to decision.

Practitioners and the academic researchers spend a great deal of time and effort focusing on strategic decisions. Every decision seems to be strategic. Although decisions related to business strategy are important, the majority of decisions, major and minor, are operational. A review of the research related to strategic decision making indicates that from the practitioner perspective, most of what is called strategic decision making is operational. When Dutton and Dukerich reveal the types of decision processes that have been included in the research, the decisions are mostly related to managing the operational issues.

The academic perspective

Decisions are not always based on logic and orderly rational thinking and analysis. Intuition and emotion, as suggested by Simon, in "Making Management Decisions: The Role of Intuition and Emotion" [5], cannot be excluded when judgment is required. The tools for rational decision making may be applicable in well-structured situations, but they seldom provide a workable decision process in which inconsistencies, incongruities, and dysfunctional data lead to several acceptable scenarios. Simon relates comments from Barnard (1938) [6]. Barnard's thesis was "that executives do not enjoy the luxury of making decisions on the basis of orderly rational analysis, but depend largely on intuitive or judgmental responses to decision-demanding situations."

With the introduction of operations research techniques and the business school emphasis on quantifying everything, an attitude that decisions could somehow be made by analysis of a set of numbers was created. That approach allows decision makers to rationalize the negative consequences of their decisions by referring to the numbers. The numbers may have lacked substance and verification, but it is always easy to counter any criticism of them by referring to market changes, new and unpredictable competitors, government regulations, the economy, and so on. Even human resource professionals began to adopt techniques that would somehow differentiate one person from another numerically.

The academic institutions have churned out an abundance of papers on the topic of strategic planning, which involves making effective strategic decisions. It is presented as a dynamic process in which organizations match business resources with the ability of their infrastructure to fulfill certain strategies. Annual strategic planning cannot be considered as dynamic. A once-a-year imposed planning system does not meet the requirements of a dynamic process. Strategy is a daily issue [7]. When strategic planning began to infiltrate the executive suites, executives began to talk about strategic decisions. Ansoff, in "Strategic Issue Management" [8], suggests that periodic strategic planning must be replaced by *real-time strategic issue management*. He traces the transitions from management systems of control to long-range planning, strategic planning, strategic issue management, and to what he describes as *surprise management*. Ansoff describes the strategic management system as a "systematic procedure for early identification and fast response to important trends and events both inside and outside an enterprise." The system includes monthly reviews of a strategic issues list and continuous surveillance of factors that affect the enterprise. Issues can be welcome or unwelcome opportunities for exploiting internal strengths or presenting threats because of internal weaknesses. Such surveillance should allow threats to be turned into opportunities. Ansoff concludes that strategic issue management offers the advantages of real-time response to new developments, quick internal reaction time, response to problems arising from any source, and a system that is not affected by organizational size or complexity. Strategy involves decision making, and the time to decision becomes an important business issue.

Decision making under conditions of economic and competitive stability may be made more readily through a structured rational process. When IBM not only controlled but directed the future of the computer industry, rationality could prevail. When Charlie Wilson, former CEO of General Motors, said that what was good for General Motors was good for the United States, it was difficult to argue against the statement at least from an economic perspective. Both companies find themselves in a different situation today. They no longer control the technologies or the markets. They live in an unstable environment. Fredrickson and Mitchell, in "Strategic Decision Processes: Comprehensiveness and

Performance in an Industry with an Unstable Environment" [9], tested the relation between comprehensiveness of strategic decision processes and performance in an industry functioning in an unstable environment. They found it to be negative. The authors, in spite of the evidence that supported their hypothesis, suggest that the study has a clear limitation related to causality. It is possible that performance precedes comprehensiveness. They note that "increased or decreased performance leads a firm to be more or less comprehensive in strategic decision making." In a related study, Fredrickson, in "The Comprehensiveness of Decision Processes: Extension, Observations, and Future Directions" [10], establishes a positive relation between comprehensiveness and performance in stable environments.

The list of papers and books on strategy and strategic decisions seems to grow each year, but it is difficult to find any emerging theory. Perhaps answers are searching for problems. Perhaps the researchers have overcomplicated the task. Perhaps there is no unified theory for decision making. Perhaps the researchers are expecting the impossible. Simon brings in the issues of intuition and emotion. Ansoff suggests strategic issue management in real time. Fredrickson and Mitchell raise the issues relative to the degree of comprehensiveness in stable and unstable environments. Others, such as O'Reilly [11], attempt to demonstrate that good information leads to good decision making. Fredrickson and Iaquinto [12] show that the decision process did not change significantly over a period of 4 to 6 years in the organizations they studied. They did find that it was significantly related to executive longevity, team continuity, and organizational size and performance. All of those conclusions are interesting, but they provide no specific direction for those attempting to reduce their time to decision. There is little evidence in the literature of how managers can improve their decision making.

Mintzberg, Raisinghani, and Theoret, in "The Structure of Unstructured Decision Processes" [13], attempt to discover how organizations go about making what they call *unstructured strategic decisions*. They use the word "strategic," but a review of the 25 decisions they studied clearly shows that strategy was not involved. The decisions are responses to the purposes and strategies of the organization; all are operational. Some examples include changing the retirement policy, acquiring a distribution agency, firing a radio announcer, purchasing new radiology equipment for a hospital, development of a new TV program, and development of a new supper club in a hotel. All those decisions are operational.

Eisenhardt, in "Making Fast Strategic Decisions in High-Velocity Environments" [14], explored some of the factors affecting the speed of strategic decision making. The research effort focuses on strategic decisions, but a study of the article shows a mix of strategic and operational decisions. As an example, much of the study focuses on the in-troduction of new products. One can argue that decisions related to new product introductions are operational, but that does not detract from the substance of the research. Experience indicates that Eisenhardt's conclusions are applicable to both strategic and operational decisions.

Eisenhardt provides a set of propositions that challenge the tradi-tional views of decision making. Her evidence suggests that fast deci-sion makers:

- Use more information than do slow decision makers.
- Develop more rather than fewer alternatives.
- Recognize that resolution of conflict is critical to decision making.
- Integrate strategic decisions and operational plans.

The research also revealed that:

- Centralized decision making is not really faster. It is a layered advice process, which emphasizes that input from experienced counselors provides a benefit.
- Fast decision making allows decision makers to keep pace with change and is linked to strong performance.
- A pattern of emotional, political, and cognitive processes is related to rapid closure.

Although the study involved eight microcomputer firms with a max-imum of 500 employees, rather than the giants of the industry, it should not be discounted as irrelevant to the corporate giants. Cor-porate megagiants are really only an aggregation of smaller entities or-ganized to meet specific objectives. Personal experience clearly rein-forces Eisenhardt's conclusions.

Prior literature and thought have emphasized that:

- A high level of comprehensiveness slows the decision process.
- Limited participation and centralized power speeds decision making.
- Conflicts or disagreements create interruptions and slow the deci-sion process.

There is no doubt that examples exist to justify all the propositions made by Eisenhardt. Those views are presented not to demonstrate a preference for one approach or the other, but to bring the critical issues in decision making to the forefront. All are most likely valid depending on how situations are described and the extent to which the activities are pursued. Research into decision processes probably defies general-ization.

In Eisenhardt's propositions, one must clearly understand what is meant by:

- More information
- More alternatives
- Extent and depth of the conflict
- Integration to what degree

There is a spectrum of each of those considerations. How much more information? How many more alternatives? Where does the conflict exist? Is it a conflict of ideas or a conflict in personalities? Does integration of strategy and operations really mean integration based on the consistent flow from business purposes, objectives, strategies, and operations activities, or does it mean the appointment of some coordinating function? Those are qualitative issues, and they must be evaluated against some norm. Considering more information and more alternatives, if they are in excess of what is required to actually reach a decision, could be costly and actually decrease the performance levels of the organization.

The practitioner's perspective

From the practitioner's perspective, every decision involves some degree of uncertainty. Comprehensiveness is a matter of judgment. What may be comprehensive to one person may not be to another. What may be comprehensive in one situation may not be in another. The difference between the information that is required and that which is available and is used within the limits of its meaning must be rationalized through logic, rationality, intuition and emotion. Judgment. That gap need not be closed completely. If logic and rationality were sufficient, executives and managers might not be necessary.

Experience has taught this author that decision processes need to be simplified. There is a preoccupation with strategic decision making. I question the viability of the concept of strategic decisions unless it is augmented by effective operational decisions. There is no doubt that strategy is important, but once that strategy has been described, its fulfillment depends on operational decisions. IBM's late entry into the personal computer business was not a strategic decision, nor was its announcement of the termination of an additional 20,000 employees in 1992. IBM could attempt to once again shape the future of the computer industry, but that is a strategy that must be augmented with operational plans. The announcement that General Motors would go through a major restructuring with termination of 70,000 employees and the closing of 21 plants in the following 4 years was not a strategic decision; it was an operational decision. Those two actions alone will not improve General Motors' competitive position. It may allow it to maintain its current market share.

Why not just focus on decisions, decisions of every possible kind? There is a continuum from major to minor decisions with an additional consideration of complex to simple. Major decisions can be simple, and those often thought of as minor may be very complex. Investing in a new plant is a major decision, but it may be made rationally with an analysis of future productivity requirements. The decision to invest in the development of a new product brings with it greater uncertainty. The work of Cooper and Kleinschmidt and others clearly shows that investment in product development requires taking significant risk. It also requires a great deal of faith in the people who will be responsible for transforming an idea into successful reality. A decision to change the culture of an organization (IBM, Kodak, General Motors) from one that is bureaucratically oriented to one that is innovative and responsive to customer needs is an operational decision: It is a decision for the survival of the organization.

As a practitioner in decision processes, some conclusions can be extracted from the generally accepted views that the effect of time to decision has on performance:

- Comprehensiveness does not necessarily slow the decision process, nor is it necessarily true that extensive inquiry slows the decision process. That depends totally on the degree of comprehensiveness. Taking it to the ultimate endpoint may be a serious mistake, but being sufficiently comprehensive in the inquiry to have the essential information required for the decision may be the right course to take. It depends on the circumstances.

- Limited participation and centralized power can accelerate the decision-making process. The participants must include those who can make a contribution to the decision. Although some will argue that those affected by the decision must also be considered, it is unlikely that such extensive involvement will occur. If for no other reason than its own self-interest, a performance-oriented management will consider the impact of its decisions on those affected by the decision. Centralized power can enhance or inhibit the decision process, depending on the literacy levels of the centralized power structure.

- Conflict and disagreements can either speed or delay the decision process. At the very least, conflicts must be brought out into the open and resolved in the predecision period. Once a decision has been made, it is too late. Postdecision conflict only detracts from pursuing the intent of the decision.

■ Real-time information from intimate knowledge of the related subject helps identify issues. Alternatives must be considered simultaneously rather than in series. Options must be evaluated. Imaginative, intuitive, experienced, and active counselors or confidants who have knowledge of the business and the industry and are capable of creating change rather than being dominated by it, provide significant benefits. Decisions must be integrated, and their interdependence must be viewed in relation to the business as a system.

Decision Models

Mintzberg [15] reflects on the three modes of decision processes: entrepreneurial, adaptive, and planning. The *entrepreneurial mode* is dominated by a search for new opportunities, and the orientation is active rather than passive. The *adaptive mode* involves division of power and is characterized by a reactive philosophy rather than a proactive one. Decisions are incremental, and they lack integration. The *planning mode* focuses on systematic analysis, based primarily on costs and benefits, and the integration of decisions. Mintzberg relates these three modes to strategic decisions, but they are really operational decisions. Each mode has its place, and most organizations will find application for all three modes. The time to decision in each mode will be quite different from that in others. In the entrepreneurial mode, the time to decision may be microseconds; in the adaptive mode, it may not even be a concern. The organization adapts. In the planning mode, although it is of utmost importance, time to decision may not be appreciated as contributing to business performance. There is no doubt that planning is a requirement, but time to decision depends on the amount of planning. Some situations require no planning; others require some degree of limited planning; and still others require extensive planning. The time to decision will be determined by the ability of management to do the right amount of planning.

There is no lack of decision models. Nutt, in "Models for Decision Making in Organizations and Some Contextual Variables Which Stipulate Optimal Use" [16], presents and critiques six decision-making models. He describes them as bureaucratic, normative decision theory, behavioral decision theory, group decision making, equilibrium-conflict resolution, and open systems decision making.

Bureaucratic decision model

The bureaucratic decision-making model involves people with competence and power as interpreters of some master plan. It involves knowledge of rules and procedures and a focus on efficiency. This approach to decision making seldom works in a dynamic situation.

Normative decision model

The normative decision theory model is characterized by

- Certainty
- An assumption that goals are known, information is available, and alternatives are available for consideration
- A mutually exclusive set of states of nature
- A probability that each state will occur
- A matrix of alternatives and environments
- A set of criteria

Normative decision theory postulates that a rational decision maker seeks to select the alternative with the maximum probability of success.

Behavioral decision model

The behavioral model suggests that a *decision space,* which considers alternatives and the states of nature, should be defined to make it workable. Before the alternatives can be developed, it is necessary to describe the acceptable states of nature. If alternatives are easy to describe, the decision maker's aspiration level rises; if not, the aspiration level declines and the decision maker redefines what may be acceptable. It is suggested that this behavioral principle of decision making makes the process more manageable.

Group decision model

Group decision making, as the name implies, means leading the group toward a decision. That approach is an outgrowth of social psychology, and no research has demonstrated that better decisions are made by its use. Nutt suggests that for decision tasks, homogeneous groups with moderate status differential are optimal and that the procedure focuses discussion and permits information and criteria to be shared by the group, which increases the possibilities of reaching a consensus. Nutt also states that: "Groups can be dysfunctional. They are costly, cantankerous, hard to dismantle, and difficult to manage." Anyone who has ever participated in extensive group activities certainly understands the circumstances that lead to dysfunctional consequences.

Equilibrium conflict resolution model

Nutt relates the equilibrium conflict resolution model as suggested by March and Simon, who postulated that decision making is stimulated by individual and/or group conflict. Uncertainty forces the decision maker into a state of conflict with either himself or the group. When the past cannot be used to make current decisions, conflict arises. Alternatives that do not have a preponderance of evidence either for or against create conflict. March and Simon suggest that group conflict is an unstable condition and that the group will seek equilibrium. This uncertainty can be reduced through problem-solving techniques, persuasion, bargaining, and politics. March and Simon failed to mention power as a means of resolving issues.

Open-system decision-making model

The open-system decision-making model assumes that reaching decisions is a complex task because certain key variables associated with the decision cannot be adequately described or understood. The required conditions are not predictable. The approach recommends a process of "adjustment" to make incremental decisions as the views of the participants are revealed. By use of this type of decision process, an attempt is made to monitor the situation until a significant request emerges. The decision process does not conform to a set plan; it adapts to the environment. It is dominated by personal or functional and group agendas, and it is based on the theory of survival of the fittest through whatever means may be necessary. Nutt mentions that some may be offended by the apparent heavy handedness, but the model does describe some processes of decision making.

Developing a strategy requires making decisions. Certain decisions must be made about the assumptions and the desirability of pursuing a particular course of action. Two additional schools of thought may shed further light on how decisions are made. Mintzberg [17,18] promotes the concept of the *emerging strategy,* which requires thinking and acting simultaneously. His thesis is that the crafting image more appropriately captures the process by which strategies come about. The structured planning image distorts the processes and sends organizations in the wrong direction. Mintzberg takes the position that "strategies are both plans for the future and patterns from the past." His metaphor of the potter suggests that "to manage strategy is to craft thought and action, control and learning, stability and change."

Ansoff [19] defends the design school approach to developing a strategy. Basically, the design school model basically relates the threats and opportunities and the key success factors along with strengths and weaknesses of the organization and its distinctive competencies in the creation of a strategy. The premises that underlie the design school model are shown in Fig. 7.1.

Mintzberg and Ackoff, in two separate articles, appear to be at opposite ends of the spectrum. Mintzberg, in "The Design School:

1. Strategy development is a conscious and controlled thought process.
2. Responsibility rests with the chief executive officer.
3. The model must be simple and informal.
4. Strategies should be unique: the best result from creative design.
5. Strategies emerge from this design process fully formulated.
6. Strategies should be explicit and, if possible, articulated.
7. Implementation demands unique, explicit, and simple strategies.

Figure 7.1 Premises underlying the design school model for formulating strategy.

Reconsidering the Basic Premises of Strategic Management" [18], criticizes the design school approach from the standpoint that it separates the formulation and implementation of strategy. The thinking and acting are viewed as two distinct activities. He basically disagrees with the seven premises of the design school. Ansoff, in a "Critique of Henry Mintzberg's 'The Design School: Reconsidering the Basic Premises of Strategic Management'" [19], responds defensively to Mintzberg's critique of the design school:

> According to Mintzberg, for all intents and purposes, all of the prescriptive schools for strategy formulation should be committed to the garbage heap of history, leaving the field to the "emerging strategy" school which he represents.

As a practitioner it is difficult to understand how intelligent and well-respected academics can conclude that they have the best process for developing strategy, or more appropriately, making strategic decisions that in reality are primarily operational. Research into the organizations Ansoff and Mintzberg consult with, and the performance of those organizations over a period of years, could be a good exploratory topic for a doctoral dissertation: The Impact of Academic Consultants on Long-Term Business Performance. Ansoff and Mintzberg should recognize that the design school of thought and the Mintzberg school of thought must be integrated. Neither ever conclusively states just what kind of strategy/operations he is concerned with. If both are dealing with corporate strategies/operations, they should go back to the drawing board for a higher degree of realism. Are they focusing on broad corporate strategy, individual division strategies of the multidivisional organizations, functional strategies, or technology strategies? Are they considering the strategies of the organization that functions solely in a local economy—a global economy? What are the differences between the decision processes in a start-up venture and a well-established organization? The strategy spectrum begins at some point either through an individual's response to achieving purposes and objectives or with some formalistic process. The strategy directs the operational decisions. It is a creative process structured no differently than the process of making operational decisions. Certainly a strategy/decision process would not begin with the idea that it should not be unique, that it should not be simple, or that it should not be explicit (three elements of the design school model). Design of a strategy through decision processes is not different from the design of a new product or process. The design begins with some limited knowledge and experience. It begins with an idea, develops into a concept, is turned inside out and upside down, and is reconfigured many times before anything useful emerges. It usually involves many operational decisions. As has been stated previously, purposes, objectives, strategies, and operations form a loop. A change in one affects the others. Before something can be molded, the potter must have some idea of the outcome. To carry Mintzberg's metaphor one step further, the clay is the design school input that emerges in some type of output through the infrastructure as a result of the resources and associated activities.

Decision Support Systems

In recent years, decision support systems have received much attention. Expert systems, artificial intelligence, real-time information systems, and simulation represent the more common tools. But it is important to understand the limitations of those tools. Most important decisions at the corporate level must be made in real time. An expert system may provide some benefit if the decision concerns the purchase of a robot or some specific piece of equipment. It may assist in the maintenance of that equipment. But can an expert system be used effectively and economically in determining what new technologies should be pursued? Can an expert system be used to resolve the issues plaguing IBM and General Motors? The tools provide opportunities to explore more alternatives, but those alternatives cannot be explored solely for the purposes of exploring still other alternatives. They require thought before action. The difficulty with most expert systems lies in the verification of the input. That input must be validated. It must go beyond asking for an opinion. Even with a simple maintenance-type expert program it is difficult to ascertain the cause of a malfunction in maintenance or manufacturing. When an automatic machine stops, does it stop because of a malfunction or because a part did not meet specifications?

Van Hee, Sommers, and Voorhoeve [20] describe a formal framework for modeling complex systems that consists of a meta-model for discrete events systems, a language based upon the meta-model, and a software environment for editing and validating system descriptions. They cite as an example the scheduling in a job shop. Jobs are entered from the sales department into the decision support system (DSS) and are divided into a set of tasks. A task is given a machine type, a standard duration, and a set of preconditions that must be met. The DSS then produces schedules for the shop floor and assigns the machines at a specific time. As the activity progresses, information is fed back to the DSS, completions are documented, machine downtime is reported, reports are prepared for the decision makers, and inputs from them are accepted. Such a system is relatively straightforward. The system is limited, and the expectations of the system are clearly described. Applying a similar system to aid in the acquisition of a new business or a new technology represents several orders of magnitude in added complexity. Decisions of that type will not be made on the basis of past performance or with old economic justification models. When an organization is breaking new ground in technologies or markets, such models can lead the decision makers down the wrong path.

Knowledge is the critical issue in decision making. Depth and breadth offer advantages, although they do not necessarily guarantee success. McGovern, Samson, and Wirth, in "Knowledge Acquisition for Intelligent Decision Systems" [21], combine decision analysis and expert systems as intelligent decision systems and a means of supporting strategic decision making. Unfortunately, the authors present only a construct of the process and suggest that although the methodology has been developed, the next step is to empirically investigate the effectiveness of the approach. Raghavan, in "A Paradigm for Active Decision Support" [22], reports on JANUS, a research prototype of a system by which the support tools are used in the decision-making process and decisions are made through collaboration between a human and the machine. Raghavan considers most DSS as passive because they represent only a tool for the decision maker, who must exploit the facilities in the decision-making process. His main concern is that DSS, as conceived, can only respond to the user's request; it cannot take the initiative. One cannot argue about the benefits that could be obtained from intelligent decision systems: The authors assume that current DSSs and expert systems provide effective decision support. The DSS must take into account such unquantifiable aspects as the action orientation (reactive, proactive, inactive) of the decision maker, specific organizational factors, the traits and characteristics, and the cognitive processing competence of the decision maker. The unquantifiable aspects will have to be included before the DSS will provide more than computerized data.

Groups often become involved in the decision-making process. Group decision-making processes as described here refer primarily to the working sessions in which acceptable solutions are developed by individuals from different disciplines and functions. Such groups can be aided by a proper DSS, but the DSS will be described differently by every organization and by every function within the organization. If an organization attempts to introduce the concept of system cycle time management, the DSS must meet the needs of all those involved in the project. Jacob and Pirkul, in "A Framework for Supporting Distributed Group Decision Making" [23], suggest that current approaches to group decision support systems (GDSS) have focused attention on facilitating group meetings. They suggest that the number of group meetings can be reduced if the participants can exchange information and expertise on a continuous basis. Their approach is to combine an organization's information system with a network of knowledge-based systems. The authors recognize that the critical factor of such a system is the knowledge base, but they provide no guidance for implementation or empirical data to justify the value-adding benefits of such a system. The authors focus on reducing the number of meetings, but caution must be exercised in using only electronic means of communications. Well-structured and disciplined meetings allow pursuing related subject matter in greater depth.

Impediments to Time to Decision

There are undoubtedly countless impediments to minimizing the time to decision. This topic has not been researched to the extent that some definitive construct emerges. At best, the input is anecdotal, but lessons can be learned from what is available in the literature. This section is a mix of personal experiences and observations as well as an attempt to stir the thinking processes of managers. Much could be accomplished if managers began to question not only their current practices in decision making but their approach to the general subject of managing.

Both time to decision and the quality of the decision are influenced by the manner in which the decisions are made. Managers have been influenced by the social psychologists who have promoted such approaches as participative management, groupthink, and consensus management. They have endorsed a kind of collegial approach to decision making. At the same time, no evidence that has ever been provided justifies such approaches in a generic sense. Claims regarding various approaches and the benefits of getting more people involved in the decision process have been made, but the impact on business performance solely related to the degree of participation has not been documented. That is not an uncommon difficulty when researchers attempt to isolate the value added by some specific program and especially when the program depends on intervention by humans. It is often just as difficult to isolate the benefits of a particular technology, a new product introduction, or plant automation. How can the benefits of the computer control system be differentiated from the equipment and the process it is controlling? That may be an academic exercise, but it is not a practical one. A control system is required; it is an essential part of the system.

The effect of participative management, groupthink, consensus management, belief structures and information processing, organizational politics, and creativity affect the time to decision. They are important considerations. There are no definitive answers, but some input provides opportunities for exploring new ways to reduce the time to decision.

Results of limited research

Hickson et al., in *Top Decisions* [24], identified nine specific impediments to the decision-making process. Figure 7.2 shows an arrangement in ascending order of factors that impede decision making. The first three are considered relatively unimportant; the last six are very important. The fact that internal and external resistance are impediments to decision processes suggests that all levels of management must review their management methods. There are other qualitative factors that the author has found through personal investigation. They include:

- The impact of what managers read. If all read the same material, they would arrive at the decision processes with the same biases.

- Lack of controversy that would force people to think deeply about their decisions.

---

1. Sequencing   Waiting to get attention to the idea or concept—getting in the attention queue
2. Coordinating   Getting the right people together
3. Timing   The day when favorable environmental, economic, and business conditions are present
4. Searching   Compiling all the data necessary for a realistic business analysis—objectives, costs, markets and all the parameters required from research, development, manufacturing, and marketing
5. Problem solving   Resolving the technical, marketing, and business issues that would in any way hinder success
6. Supplying   Awaiting the availability of adequate resources
7. Recycling   Reconsidering and awaiting decisions on what has already transpired
8. Internal resistance   Counteracting active internal opposition
9. External resistance   Thwarting and awaiting any interest or opposition that delays the decision

---

Figure 7.2 Reasons given by respondents for delays in decisions. They are arranged in ascending order of impeding decision making.

- Extrapolation of yesterday's results to describe the future.
- The crisis orientation of managers.
- Levels of illiteracy about the elements involved in a particular decision.

Participative management

The principles of participative management are well known [25,26]. As the term implies, employees are supposed to want to participate in the decision process. There is no evidence that isolates the real benefits of participative management. There is no evidence that participative management is any better than authoritarian management. One can reason that if employees are given an opportunity to participate, performance should improve. However, if an organization is managed according to the three E's, participation is not a choice, it is an expectation of the participants. Much of the pro and con argument arises when attempts are made to describe it. Participative management is not managing by majority vote; it requires involving any individual who can provide a productive input to the process. In the final analysis, someone will have to reconcile all the different requirements and reach some form of compromise. A decision involving inputs from scientists and engineers of many different disciplines will not be reached by continuously meeting to resolve some form of appropriate compromise. At the same time, many minor decisions can be passed down to lower levels in the organization. However, allowing the newly minted engineer to determine the particular technology in a major program could be a serious mistake.

Groupthink

Janis, in *Groupthink* [27], describes the antecedent conditions and the observable consequences of groupthink. Groupthink refers to the tendency of cohesive groups to become so concerned about group solidarity that they fail to critically and realistically evaluate their decisions and antecedent assumptions. According to Janis, that condition occurs when the antecedent conditions (insulation of the group, homogeneity of members, lack of impartial leadership, lack of any norms relative to procedure) and high levels of cohesiveness are present. Janis's study involved 16 cases: 7 included graduate and undergraduate students, 6 related to sociopolitical situations, 1 had to do with content analysis, and 2 were business-related—one to the development of the Edsel and the other to autonomous work-group processes.

Park, in "A Review on the Research on Groupthink" [28], believes that the concept has considerable practical value. He states that when decision-making groups fully understand the groupthink model and try not to become victims of groupthink, the concept should be of great value to certain types of organizations. Empirical studies of groupthink may not be possible because the study would have to be performed in real time. Little benefit would be achieved in questioning those involved in the process after the decision was made. One can argue the issue related to the impact of cohesiveness, but from a practical perspective, cohesiveness seldom allows the devil's advocate to enter the scene. Cohesiveness usually involves maintaining a predetermined path or plan or methodology. It is equivalent to maintaining the status quo and not rocking the boat. Any activity or thought that would tend to disrupt the cohesiveness of the group is looked on as an unnecessary intrusion.

The report on the space shuttle Challenger accident in January 1986 revealed much information as well as controversy about the management and decision processes of NASA. Esser and Lindoerfer, in "Groupthink and the Space Shuttle Challenger Accident: Toward a Quantitative Case Analysis" [29], provide the results of their research as a result of coding, positive and negative, instances of the observable antecedents (as stated previously by Janis) and consequences of groupthink. They conclude that the decision to launch the Challenger involved groupthink. Research by Callaway, Marriott, and Esser [30] has also differentiated the results of groupthink depending upon the types of individuals in the process. They found that groups composed of highly dominant participants made better decisions, exhibited lower anxiety, and made more statements of agreement and disagreement during the decision process than did groups composed of low-dominance individuals.

Consensus management

Reaching a consensus is obviously a requirement in business. Organizations could not function without some level of consensus, whether that consensus involves any of the activities related to product design or to managing the business. However, the need for consensus and the development of consensus must be integrated. Consensus, yes, but how? Is consensus accomplished through in-depth understanding, or is it accomplished through coercion? Does it become personal? Does it involve elements of groupthink—cohesiveness?

Bourgeois [31] studied 12 nondiversified companies relative to management consensus related to corporate objectives and the means for attaining those objectives. The results indicated that (1) consensus of means yields higher performance than disagreement on means, (2) allowing disagreement on less tangible goals tends to be associated with better performance, and (3) the worst performance results not from goals agreement, but from means disagreement. Although the study was broadly based, 9 of the 12 organizations involved high technology. Bourgeois [32] cites his dissertation at the University of Washington (1978), in which he states that goals disagreement correlated positively with economic performance when managers had an accurate perception of uncertainty. DeWoot, Heyvaert, and Martou, in a study of 168 Belgian firms [33], found that the more successful firms were characterized by frequent disagreement on the alternative means to achieve the goals of *technological innovation*. Grinyer and Norburn [34] investigated the relation between strategy-making practices and performance of 21 British firms in 13 industries. They found no correlation between goal consensus and performance, but among the six best performers, they found that goal disagreement related positively to performance.

Belief structures and information processing

Problem identification and definition affect the time to decision. Past experience, bias, opinion presented as fact, the organizational infrastructure, and so on, determine the level of integrity of the problem description. In many situations, cause and effect are reversed because of the manner in which the problem is structured. As an example, consider the situation in which sales of a particular product continue to show significant decreases with each quarter's report. Someone may offer the comment that the reason for the reduced sales is product quality, so the assumption is made that quality is the problem. Quality is really not the problem; quality is the answer. The elements of the infrastructure that allow the manufacture of a product that does not meet the marketplace requirements is the real problem. Focusing on a specific quality program will have little impact until the antecedent conditions of the infrastructure are strengthened in such a way as to allow a quality product to be manufactured. A similar statement applies to technology: By itself, technology is seldom the real problem. A technology problem most often stems from technical incompetence and managerial illiteracy in technology. The manner in which individuals process information and individuals perceive the structures affect the way in which problems will be perceived and resolved.

Walsh, in "Selectivity and Selective Perception: An Investigation of Manager's Belief Structures and Information Processing" [35], reexamines the results of Dearborn and Simon [36] regarding the manner in which managers use selective perception in processing information in decision making. Dearborn and Simon concluded that managers develop a viewpoint over time that is consistent with the departmental or functional goals and activities in which they participated. Walsh examined four hypotheses:

1. A manager with a belief structure arising from working in a single functional domain is likely to have worked in that domain.

2. In decision situations that are not adequately structured, (a) managers are most likely to *identify problems* from the same functional domain as the content of their belief structure, (b) managers are likely to *use only the information* that came from the same functional domain as the content of their belief structure, and (c) managers will *seek additional information* from the same functional domain as the content of their belief structure.

Basically, Walsh attempts to disprove the conclusions of Dearborn and Simon and show that managers do not "suffer from impoverished world views or parochial information use."

Organizational researchers have assumed that the basic belief structures of managers, when the decisions are not adequately structured, have detrimental effects on decision making. Walsh attempts to disprove these assumptions by doing research in a vacuum with an antiseptic approach. He does not deal in the real world. His research is based on data from 121 midcareer managers in a part-time executive MBA program. It was a study in beliefs, but not beliefs related to decision making. No connection was made between beliefs and practice. The participants never became involved in decision making in Walsh's study. Espousing a particular belief is not the same as implementing it.

Organizational politics and creativity

Organizational politics and individual creativity interact in the time to decision considerations. Unfortunately, organizational politics and power are viewed as evil and individual creativity as good. But politics need not be evil, and creativity may not always make a contribution in the decision process. Decision making is situational. Whenever two individuals communicate, the elements of politics and power become apparent. If there are differences, one will attempt to convince or coerce the other; if there is agreement, both will attempt to convince or coerce others to accept their position. There is no doubt that politics often consumes time, restricts or limits information, creates communication barriers, and generates dissatisfaction. But the opposite is true as well. The right type of politics can cut through a malaise of decision processes, acquire information that may have been hidden or restricted, open up communication, and motivate a group because of the levels of accomplishment.

The foregoing comments apply to the scientific and engineering community also. Research, design, development, manufacturing, and marketing are not without their politics. Withholding information and lack of communication are major problems in the scientific and engineering communities. All functional groups or teams accomplish their goals through various types of political and power alliances. It is important for executives and managers to recognize when politics and power become a destructive force. Although I cannot provide any major research results, my personal experiences indicate that politics and power play a major role in decision making and are a means of reaching better and more appropriate decisions. One major condition must exist: The politics must be based on facts that are often in opposition but are resolved by means of an appropriate political and power structure. All the essential information should be exposed for the necessary scrutiny of validity and importance. Badawy [37], in considering some of the elements that are required to succeed in technical management, states: "Informal political systems and power structures are typical of organizational life. Learning how to handle organizational power and politics and survive is a crucial interpersonal skill every manager should have."

Some level of creativity is essential in all decision processes. That is certainly true of decisions related to technologies and markets. If the decision is to have any advantage, that available data and information must be looked at differently. The creative spirit will have to be released if something new is to be extracted from a common database. Scientific papers can be reviewed and evaluated, but the process must include the stimulus for thinking differently about particular subject matter. It requires putting information together in a different way and drawing new conclusions. Summers and White, in "Creativity Techniques: Toward Improvement of the Decision Process" [38], review the various creative techniques such as brainstorming and nominal group technique. These techniques, although more applicable to problem solving, can provide additional input to the decision makers. They conclude that the output emanating from such groups can be a major source of new insights and information that have the potential to improve the quality of the decision. If the assumption is accepted that most decisions include some elements of organizational politics and power, then creativity, willingness to raise the controversial issues, exploring the reasons for dissenting views, and so on can be a major force in using politics and power as a positive source that enhances performance.

Factors Affecting Time to Decision

The factors affecting the time to decision can be argued pro and con, but eventually the decision process depends on how the participants view a particular situation and from which perspective. Within system, product, and project cycle time many decisions that affect total time, timing, and cycle duration are made. Identifying that lost time and then adapting the decision processes provides many opportunities for improving a business unit's performance. The business, products, technology, and marketing activities account for most of the lost time attributed to time to decision.

Business

How many hours of lost time are generated in an organization because of the lack of a timely decision? There is no conclusive research to provide a quantitative figure that has meaning across a broad spectrum of industries or businesses. But executives must assess the situation in their own organizations. As a rule of thumb, a delay in a major decision from the CEO's office costs at least one hour per day per employee involved. If 1000 people are involved, a minimum of 1000 hours of work are lost the first day. That lost time continues to impair productivity until the participants either recognize the importance and validity of the decision, faintheartedly accept it, or choose to ignore it. It makes no difference whether those involved are part of management, part of the professional community, the secretarial or support staff, or the production people. Those discussions commence with entry to the workplace, and they continue as the primary topic of conversation at the vending center that extends the break period, continues at each interchange between individuals, arises as the first topic of discussion at the beginning of every meeting, enters the rumor mill that generates additional lost time, and whirls around in the minds of each employee and thus affects levels of concentration. Uncertainty consumes time until the uncertainty has been removed. That scenario may not please management, but those who do not concur with it need only test the scenario at their own levels the next time they are affected by a delayed decision.

Some examples may help to show the importance of time to decision as it relates to the business level:

1. For some unknown reason, organizations make structural changes without making the necessary new appointments. The business press continually announces promotions or new assignments without drawing attention to the vacancies created. Some vacancies continue for many months; in the interim, the group slowly loses its direction.

2. Delays in approving projects of any type and for any functional group. There is no doubt that megacorporations must control the purse strings to some extent. But the number of desks at which that paperwork must stop for a rubber stamp delays decisions.

3. Delays in issuing policies related to any aspect of the business translate into wasted time. Whether the policies involve human resource interests such as sexual harassment and personnel evaluation, technical agreements, quality, automation, ethics, and so on, time will be consumed without any productive effort from the time the rumor is born until formal policy is issued.

4. Delays in the decisions involving the general operations of the business that may include decisions related to acquiring or divesting businesses, technologies, and so on, entering into new businesses, expanding current businesses, entering new geographical markets or proceeding with global expansion, closing down nonproductive operations, introducing a new management information system, and even down to such simple decisions as changing the color of paint in the offices.

5. A timely topic involves early retirement programs or some form of termination incentive. Why do organizations implement those programs? One part of the answer is simple, the other more complex. The simple part is that few decisions must be made about who should be released. The potential early retirees have sole responsibility. The complex part is that some organizations have delayed the termination of marginal employees for many years and now the costs continue to increase with no value-added benefit from their activities. Performance expectations never receive a priority in managing the human resources.

The complexity of the process arises from the fact that decisions are seldom made in isolation of past or future decisions. Acquiring or divesting of a business or a technology will affect all of the technology- and marketing-related functions. Making a single decision presents few problems. When managers begin to consider their about to be made decision in relation to the business system, procrastination raises its ugly head and significantly delays if not paralyzes the decision process. Managers are supposedly paid to make decisions. Unfortunately, too few managers are risk-oriented and can function effectively in a climate that is dominated by uncertainty. All the delays cause lost productivity. Every organization must make up its own list and begin by focusing on the decisions that have the greatest impact.

Products

There is little argument that time to decision, relative to continuing current products or introducing new products, poses challenges to managers. The product decisions begin with preliminary specifications and continue throughout the product life cycle. The following examples illustrate the impact of the time to decision as related to products:

1.   A decision on strategy prior to embarking on a new product venture is obviously necessary. Does the business choose to be a leader or a follower? Does it function within a set of narrow guidelines that describe its product portfolio? Does it choose to focus attention on specific types of products? Does its strategy allow the participants to become involved in new to the market product or in breakthrough products? Decisions must be made in each of those areas, and the time to decision is important. Decisions delayed at this point are costly.

2.   What are the specifications for a new-to-the-market product? There are often thousands of decisions of varying importance that must be made. They must be made somehow, and agreement reached on their consequences must be reached. Those decisions relate to form and function, and they also take into account output, reliability, beneficial features, energy consumption, environmental considerations, and the ease of operation. Each of those will involve certain tradeoffs by the participants and require timely actions for optimizing cycle time.

3.   Introduction of new products must be guided by some formal procedure; the helter-skelter approach does not work. Implementation of that process involves timely decisions. Some flexibility is essential because there are times when, for justifiable reasons, certain parts of the procedure may be eliminated. But that elimination must be recognized and its consequences must be evaluated before the decision to bypass it is made.

4.   The selection of which products to pursue involves timely decisions. The screening process involves many decisions that are related not just to the product but to the markets, the distribution system, the financial requirements, the level of technology, and so on. Most of those decisions are not either-or, black or white; they are in the gray zone. The decision to proceed after the feasibility study has been completed cannot wait for several months. If it is delayed, additional time is then needed to reorient the participants. If the participants have been dispersed, additional time is lost.

5.   The time and effort required to develop a business plan for a new product involves using the input from many people as well as the business resources. The business plan must include more than the sales and profit projections for a 5-year period. Business plans go through many iterations because managers do not identify the key issues at the beginning of the process. They focus on the quantitative exclusively and even quantify the qualitative data without any verification. The activities related to preparing such a business plan require many hours of study, analysis, and synthesis. Time is involved in doing the up-front preparation, but usually it is minimal as compared to the time lost if the discussion and approval of that plan requires continuous modification. Approval of the plan must be scheduled well in advance of the

completion date. If one of the key decision makers goes on vacation when the plan is completed and no action takes place for a month or more, costs continue to rise. The individuals developing the plan may use the time to fine-tune it, but that is most likely an unnecessary step. The plan in all its aspects will not be within 10 percent of any one financial line, so why waste the time? The decision makers must be available when needed; otherwise, time is wasted.

Technology

Decisions related to technology involve products and markets. The first prerequisite is that management fully understand that technology is a business and not a technical issue. Until business and technology achieve congruency, technology will continue to be misapplied and development of elegant technology will take precedence over developing reliable products. Technology cuts across research, development, manufacturing, marketing, and all the administrative functions. Those are basically product and process technologies. Information technology encompasses all the business functions, but timely decisions are essential not only within each of the functions but among all of the functions. Some examples will demonstrate the impact on performance.

1. It is absolutely essential that those who make decisions related to technology have some minimum level of technological literacy in order to speed the decision process. Fear in decision making arises from lack of understanding. Executives who do not understand the limitation of an organization's current technology base and are illiterate in the technologies being developed beyond the state of the art lengthen the time to decision when technology enters the decision process. They must be educated at every step. That does not imply education in depth, but some understanding of the fundamentals and the potential of the technology for affecting future performance is necessary. Their questions are often endless and delay a decision. Lack of understanding by the decision makers leads to requesting more information, suggesting and evaluating new alternatives, and introducing other intentional or nonintentional delays. The greatest difficulties arise when the executives who are responsible for managing technology raise those issues. They cannot see the big technological picture because they no longer keep abreast of the advances in technology. In the process, time to decision is increased and resources are wasted.

2. Researchers must also consider the timeliness of their decisions relative to technology matters. The time to decision for researchers involves more than the strictly technical decisions. The decisions relative to material purchases, the need for services of others, and the next step

if the results of the current experiment are not successful must be made in a timely manner. The technical decisions are important, but the ancillary decisions consume the time of others because of the panic approach often associated with them. In the interim, the researcher's primary activities are sidetracked and are accompanied by a lack of productivity. When lack of concern for the time to decision is associated with functional groups that rely on research for timely performance, the wasted time and the associated cost increase geometrically.

3. The development function adds cost logarithmically. Research, as contrasted to development, is the least expensive. It involves new discoveries and often new inventions. But development is generally a long and costly adventure. Much of that cost comes from disregarding the time to decision. As an example, it is virtually impossible to retrace the development of xerography and determine the decision-making processes that took place or even if there was any consideration of decision making as a formal process. The fact remains that it took 22 years to go from concept to demonstration of an acceptable and marketable rotary drum copier. A similar statement can be made about most major product breakthroughs. But decisions in the design and development phase of a product involve the total spectrum of decision complexity. All the decisions are important. As an example, the decision to limit the number and type of fasteners in an assembly has a significant impact on design, production, maintenance spare parts, and final manufacturing cost. It affects purchasing, production, inventory control, and so on. It may be a relatively minor decision, but it must be made before the design phase begins, not at some intermediate point when large numbers of engineering drawings will have to be changed. The decision to use a modular approach in the product design to provide easy maintenance must be made in a timely manner. Decisions relative to materials, tooling, and processing cannot wait until manufacturing takes over to produce the product. Time to decision in development affects the decision stream to manufacturing. Delays in the decision process add cost and lengthen cycle time.

4. Establishing basic design criteria for a new product or process would appear to be a fundamental practice or a given. Unfortunately, too may designs are accomplished by using an ad hoc approach. Designs that involve many different participants need a set of design ground rules if at some time in the future those parts or subassemblies are to be brought together into a workable product or process without a major redesign. The number of change orders to be issued in the future depends on how clearly the ground rules are communicated and implemented. However, establishing the design criteria requires many different decisions beyond those made in the design department.

During a project review, the project manager inquired about the latest design review that took place with one of the subcontractors. The question related to the approval of the design. The design engineer reported that the design had been approved, but with a very large number of deviations. In other words, to proceed with the tooling and so on, the designer allowed the deviations. There is no doubt that some deviations in complex equipment or products may be required. However, the number of allowed deviations is a good indicator to management about the capability of those who developed the performance criteria. It is a measure of the quality of the up-front effort and the decision processes. In this case, taking more time to reach a decision on the performance criteria would have been a better approach. Every change and the acceptance of deviations from specification create additional problems and cost downstream, and they lengthen total cycle time because of decisions made without sufficient forethought.

Marketing

Markets and market share are won or lost depending upon the time to decision. IBM managed to capture a major share of the personal computer market only after an infusion of vast resources. It would be interesting to explore the decision processes used by IBM to stay out of the personal computer market as contrasted to the decision processes that forced the company into it. Perhaps the decision to stay out was due to neglect, hubris, and arrogance and the decision to enter was due to panic. There is no doubt that decisions relative to marketing share the same risks and uncertainties as those related to technology. Certainty does not exist. Time to decision, as with business, products, and technology, affects the final cost of entry.

Marketing decisions, like all other decisions, must be made in a timely manner. Some of the decisions are made in the product conceptual stage and continue until the product reaches the marketplace, where its true value is finally measured. There is no one place in that continuum where one decision can be considered more important than another if it affects cycle time. Yet the time to decision affects the cycle time and the total number of hours spent on meeting the project objectives. The following examples related to the five stages of a product's life cycle, namely, concept, feasibility, realization, commercialization, and withdrawal, illustrate the significance of time to decision.

Conceptual stage. During the conceptual stage of a new product, marketers must contribute their knowledge to determining the basic parameters around which the product will be developed. Although the technology requirements must be defined by research and development, what is expected of that new product is a marketing decision. When a product concept originates in the technology functions, marketing must also provide the necessary input. At this stage the investments may be relatively small, but the integrity of the information must be validated. Late decisions or decisions made in a perfunctory or casual manner cause major future difficulties. It is as a result of such actions that concepts are scrapped, projects are canceled, and new processes are begun. This type of action not only lengthens the cycle time but increases the total time and thus the cost.

Feasibility stage. The feasibility stage is equally important. Market feasibility is not a solitary decision; it is a continuum of decisions. Many market-related decisions, including verification of customer needs, potential sales volume, selling price, product synergy, potential for expansion geographically, and distribution system, now enter the new-product process. A decision to proceed with the development cannot be made until all the market-related decisions are finalized within the limits of the available information. The process is iterative. It is at this feasibility stage that management must make a yes or no decision and major business resources must be committed. Delays in the market-related decisions add cost not only to marketing but to all the related functions. As with the conceptual stage, a casual response to the market-related issues will add time and cost in the future.

Realization phase. The realization phase of a new-product project that includes research, development, design, demonstration, and a business plan needs timely decisions from marketing. A delay of 1 day can be easily multiplied to 10 days or more or, on occasion, more than 100 days. It depends on how many downstream decisions are delayed. Unfortunately, some activities must follow a series pattern. Manufacturing cannot determine factory cost without a market forecast of the volume and the volume mix. Obviously, manufacturing could prepare a matrix of cost curves and then make a simple choice when marketing presents its forecast. But that type of activity involves additional time for vendors to prepare quotations, for manufacturing to plan all its resources under a multiplicity of scenarios, for selecting processes based on volume requirements, and so on. The approach can be an unproductive additional burden that manufacturing must avoid, especially on complex products or equipment. It does not add any value. That time could be spent more productively on some other activity. That is not to suggest that, at times, such approaches may not be necessary, but they cannot rule the daily activities.

During this phase, marketing must also make many decisions relative to the sale and distribution of the product. The decisions relate to product strategy, the market segment, current market position, pricing, distribution, the customer base, the marketing plan, advertising, the key success factors, inventories of current products to be withdrawn, and so on. A late decision in any one of those activities increases the time from concept to product introduction. The input from marketing to the business plan must be based on validated information. Optimistic sales forecasting appears to be a malady in marketing. During the review of the go-ahead for a new product, the individual presenting the marketing plan was asked to comment on the sources of the new sales. Who were some of the primary customers by name, for how many units, and on what kind of delivery schedule? Unfortunately, the response was not very convincing. Such fundamental information is important and should be verified before major investments are considered. Here is a case of the project leader asking for a major investment in the commercialization phase, and marketing had no idea of the size of the customer base. Generalization is insufficient when sales are being projected. That situation occurs on a daily basis and wastes hours of time that could be dedicated to more productive activities. Forecasting has not yet reached the point of being a science, but that does not prevent the sponsors of new products from doing their homework.

Commercialization phase. There are no fewer decisions in the commercialization phase. If this phase is approved, costs escalate geometrically. The commercialization phase includes such activities as preproduction models, internal and field testing, market evaluation, production and production scale-up, selling the product, and customer service. All those decisions are interconnected and reflect both backward and forward into the system. They are not independent of one another. As an example, production of products or equipment that will meet customer requirements of performance and quality depends on the decisions made during the period of internal and market testing and the market evaluation of the product. The decisions made in the production scale-up of a product or process will affect the physical distribution and the customer service. As a project reaches the production stages, time becomes the critical factor. It is at this time that decisions determine whether the product will reach the market as planned and meet customer requirements. In this example, the time to decision and the quality of the decision determine the cycle time.

Withdrawal phase. The withdrawal phase of a product has been included because its consequences are often forgotten. That is especially true of replacement products based on new principles. A late decision to withdraw a product from the market because of gradual obsolescence results in nonsalable or distressed price inventory, potential customer dissatisfaction, and a waste of business resources. On a tour through a warehouse with a client, a consultant asked about some of the dated inventory. The response attempted to justify maintaining the inventory on the corporate books at full value, since it would have a potential benefit in the future. That future never arrived because of cash flow problems. Inventory that cannot be sold is probably worth only its scrap value. It does not provide cash flow. Obsoleting products and replacing them with new products requires early decisions to prevent building excess inventories that will not be sold before the new product is introduced. In this case, time is of the essence and the time to decision makes the difference between success and total failure.

Summary

As applied to quality, zero defects is an ideal that is not achievable, but it provides a framework for continuous improvement. The concept of zero defects can develop a new mindset that emphasizes the need for every individual to try to reduce defects within the total business system. Defects can approach zero. However, the concept of zero defects applies not only to *things* but also to *decisions*. Manufacturing organizations generate a continuous stream of decisions every day that must be integrated with the business structure. Most organizations are not making jelly beans or gum drops—products such that the same ingredients and processes produce an acceptable output with wide tolerances of color, shape, and taste. The demands for improved performance will never stop. Some individual, somewhere, will always be trying to stress the current system with a new idea. Zero time to decision is an impossibility, but, like zero defects related to quality, it will be approached only by understanding the parameters of decision making and a consistent effort to reduce the negative effects of extended time to decision. Managers must determine the impact of the wasted hours caused by waiting for a decision. As was mentioned previously, a delayed decision at the executive level generates a lost time equivalent of an hour per day per employee affected by the decision.

Decision making is not a simple process. Businesses must understand and deal with the impact of time to decision on performance. In summary:

- The research on decision making is inconclusive because most of it is anecdotal and after the fact rather than in real time. Results are history, and history is subject to interpretation and has various levels of integrity.

- Decision models provide a perspective, but seldom is any decision made by using a single model. This author's conclusions are that decisions often are a combination of bureaucratic, normative, behavioral, group, equilibrium-conflict, and open system. Then the design school and Mintzberg approaches can be added for good measure.

- Decision support systems are important, but they must be in real time and the integrity of the input must be verified. The Frito-Lay DSS, as a substitute for an immediate response to the appropriate individual by telephone, may be a misguided use of information technology.

- The impediments must be understood as applied to a specific organization in its specific business environment. Good decisions can be achieved only if free and open discussion by those able to contribute to the process allows presentation of details often deemed irrelevant but actually important. Consensus cannot be limited to objectives. It must include the means. Disagreement is not something that should be shunned; it is an important factor in decision making.

- Some examples of extended time to decision as related to business, technology, product development, and marketing have been included. All interact and affect each other. The factors affecting time to decision cannot be viewed from a narrow functional perspective; they must be seen in the context of the business. Every organization must determine the factors that affect its operations.

References

1. H. Mintzberg, *The Structuring of Organizations,* Prentice-Hall, Englewood Cliffs, NJ, 1979.
2. C. Lindblom, "The Science of 'Muddling Through'," *Public Administration Review,* vol. 19, no. 2, 1959, pp. 79–88.
3. D. J. Hickson, R. J. Butler, D. Cray, G. R. Mallory, and D. C. Wilson, *Top Decisions,* Jossey-Bass Publishers, San Francisco, 1986, pp.96–107.
4. J. E. Dutton and J. M. Dukerich, "Keeping an Eye on the Mirror: Image and Identity in Organizational Adaptation," *Academy of Management J.,* vol. 34, no. 3, 1991, pp. 517–554.
5. H. A. Simon, "Making Management Decisions: The Role of Intuition and Emotion," *Academy of Management Executive,* February 1987, pp. 57–64.
6. C. I. Barnard, *The Function of the Executive,* Harvard University Press, Cambridge, MA., 1938, includes the essay on logical and nonlogical decision processes.
7. G. H. Gaynor, *Achieving the Competitive Edge Through Integrated Technology Management,* McGraw-Hill, New York, 1991, pp. 69–86.
8. H. I. Ansoff, "Strategic Issue Management," *Strategic Management J.,* vol. 1, no. 2, 1980, pp.131–148.
9. J. W. Fredrickson and T. R. Mitchell, "Strategic Decision Processes: Comprehensiveness and Performance in an Industry with an Unstable Environment," *Academy of Management J.,* vol. 27, no. 2, 1984, pp. 399–423.
10. J. W. Fredrickson, "The Comprehensiveness of Strategic Decision Processes," *Academy of Management J.,* vol. 27, no. 3, 1984, pp. 445–466.
11. C. A. O'Reilly III, "Variations in Decision Maker's Use of Information Resources: The Impact of Quality and Accessibility of Information," *Academy of Management J.,* vol. 25, no. 4, 1982, pp. 756–771.
12. J. W. Fredrickson and A. I. Iaquinto, "Incremental Change, Its Correlates, and the Comprehensiveness of Strategic Decision Processes," *Academy of Management Proc.,* 1987, pp. 26–30.
13. H. Mintzberg, D. Raisinghani, and A. Theoret, "The Structure of Unstructured Decision Processes," *Administrative Science Quarterly,* vol. 21, June 1976, pp. 246–275.

14. K. M. Eisenhardt, "Making Fast Strategic Decisions in High-Velocity Environments," *Academy of Management J.,* vol. 32, no. 3, 1989, pp. 543–576.
15. H. Mintzberg, "Strategy-Making in Three Modes," *California Manage. Rev.,* vol. 16, no. 2, 1973, pp. 44–53.
16. P. C. Nutt, "Models for Decision Making in Organizations and Some Contextual Variables Which Stipulate Optimal Use," *Academy of Management Rev.,* April 1976, pp. 84–97.
17. H. Mintzberg, "The Design School: Reconsidering the Basic Premises of Strategic Management," *Strategic Management J.,* vol. 11, 1990, pp.171–195.
18. H. Mintzberg, "Crafting Strategy," *Harvard Bus. Rev.,* July-August 1987, pp. 66–75.
19. H. I. Ansoff, "Critique of Henry Mintzberg's 'The Design School: Reconsidering the Basic Premises of Strategic Management'," *Strategic Management J.,* vol. 12, 1991, pp. 449–461.
20. K. M. van Hee, L. J. Somers, and M. Voorhoeve, "A Modeling Environment for Decision Support Systems," *Decision Support Systems,* vol. 7, 1991, pp. 241–251.
21. J. McGovern, D. Samson, and A. Wirth, "Knowledge Acquisition for Intelligent Decision Systems," *Decision Support Systems,* vol. 7, 1991, pp. 263–272.
22. S. A. Raghavan, "A Paradigm for Active Decision Support," *Decision Support Systems,* vol. 7, 1991, pp. 379–395.
23. V. S. Jacob and H. Pirkul, "A Framework for Supporting Distributed Group Decision Making," *Decision Support Systems,* vol. 8, 1992, pp. 17–28.
24. D. J. Hickson, R. J. Butler, D. Cray, G. R. Mallory, and D. C. Wilson, *Top Decisions,* Jossey-Bass Publishers, San Francisco, 1986, pp.108–127.
25. G. H. Gaynor, *Achieving the Competitive Edge through Integrated Technology Management,* McGraw-Hill, New York, 1991, pp. 43–63.
26. D. McGregor, *The Human Side of Enterprise,* McGraw-Hill, New York, 1960.
27. I. L. Janis, *Group-Think,* Houghton Mifflin, Boston, 1983, pp. 242–260.
28. W. Park, "A Review of Research on Groupthink," *J. of Behavioral Decision Making,* vol. 3, 1990, pp. 229–245.
29. J. K. Esser and J. E. Lindoerfer, "Groupthink and the Space Shuttle Challenger Accident: Toward a Quantitative Case Analysis," *J. of Behavioral Decision Making,* vol. 2, 1989, pp. 167–177.
30. M. R. Callaway, R. G. Marriott, and J. K. Esser, "Effects of Dominance on Group Decision Making: Toward a Stress-Reduction Explanation of Groupthink," *J. of Personality and Social Psychology,* 1985, pp. 949–952.
31. L. G. Bourgeois III, "Performance and Consensus," *Strategic Management J.,* vol. 1, 1980, pp. 227–248.
32. L. G. Bourgeois III, "Environment, Strategy and Economic Performance: A Conceptual and Empirical Exploration," *unpublished doctoral dissertation,* University of Washington, 1978.
33. P. DeWoot, H. Heyvaert, and F. Martou, "Strategic Management: An Empirical Study of 168 Belgian Firms," *Intl. Manage. and Organization,* vol. 7, 1977–78, pp. 60–75.
34. P. H. Grinyer and D. Norburn, "Planning for Existing Markets: An Empirical Study," *Intl. Studies of Manage. and Organization,* September 1960, pp. 257–278.
35. J. P. Walsh, "Selectivity and Selective Perception: An Investigation of Manager's Belief Structures and Information Processing," *Academy of Management J.,* vol. 31, no. 4, 1988, pp. 873–896.
36. D. C. Dearborn and H. A. Simon, "Selective Perception: A Note on the Department Identification of Executives," *Sociometry,* vol. 21, 1958, pp. 140–144.
37. M. K. Badawy, *Developing Managerial Skills in Engineers and Scientists,* Van Nostrand Reinhold, New York, 1982, p.10.
38. I. Summers and D. E. White, "Creativity Techniques: Toward Improvement of the Decision Process," *Academy of Management Review,* April 1986, pp. 99–107.

Foundations for Exploiting Cycle Time

System cycle time management (SCTM) requires knowledge and practice in the fundamentals of managing. Part 1 considered the limitations of describing cycle time in relation to the various approaches to concurrent engineering. It brought into perspective the need to view cycle time in relation to time and timing and gave the reader an opportunity to focus attention on the sources of wasted or poorly used time. Part 2 dissected cycle time into business cycle time, product cycle time, project cycle time, and time to decision in order to understand the interrelations and the impact on business performance. Part 3 focuses attention on four major factors that affect management practices associated with managing cycle time.

Much of an organization's effort results in action without adding value. Chapter 8, Focusing the Organization on Value-Adding Activities, raises the principal issues involved in managing in a way that every activity adds value in some manner. Managing cycle time must direct attention to the three E's throughout business-related internal and external activities. There are no consistent measures of performance for research, development, manufacturing, and marketing. Each has its own driving force as currently managed and measured. Those driving forces and those performance measures must be reconciled. There is really only one measure of performance.

Chapter 9, Barriers to Exploiting Cycle Time, raises some of the issues that prevent organizations from gaining the total

*benefits from managing for improved cycle time. Those
barriers are real, and they reside in the infrastructure of the
organization. Experience teaches that projects seldom fail
because of technological gaps. They fail because of the actions
of people at all levels in the organization beginning with the
CEO. Those issues are brought to the attention of the reader
through examples taken from current business operational
practice. Barriers to effective performance are real, and they
call for the attention of management at all levels.*

*Chapter 10, Implementing a Responsive Management
System, considers the major issues involved in bringing system
cycle time management into being. New approaches require
new practices. The organizational dynamics, the paradox
associated with information management, the discontinuities
that are always lurking somewhere on the periphery, the
measurement of individual and group performance, and
organizational learning are some of the concepts that
determine the response time of the system. All are
interdependent and either help or hinder in optimizing cycle
time.*

*There may be a price that must be paid for accelerating cycle
time. Chapter 11, Costs of Accelerating Cycle Time, brings the
reader to the realities of speed to market. Doing things fast
may be a costly adventure, especially if the speed replaces
thinking and doing the up-front work. All investment in a
profit-making organization must be evaluated and justified
either quantitatively or qualitatively. No exceptions. The
chapter raises the issues related to the time-cost relationship.
The relations of activities, process, and results must be
considered as a unit. Focusing attention on accelerating cycle
time may or may not provide benefits.*

Focusing the Organization on Value-Adding Activities

System cycle time management (SCTM) requires that managers emphasize value-adding activities throughout the organization. No function or activity can be eliminated from the process. It is important, however, to describe what the organization considers to be value adding. The descriptions of research and manufacturing could be quite different. In a discussion with researchers, it became clear that they misunderstood the concept as applied to their activities. Such slogans as "do it right the first time," when used in manufacturing operations, must be interpreted quite differently from their use in research. The first experiment, if done the right way the first time, may not necessarily yield the expected results. The world of research involves searching for the unknown. But value adding involves the total organization. As individuals become more closely allied with the implementation process, they more easily see the value of their contribution to the business.

Chapter 8 raises the issues related to adding value throughout the business enterprise. Adding value requires a system perspective, the integration of functions that often tend toward isolation, and capitalizing on the opportunities for leveraging the organization's resources. The discussion focuses attention on:

- Value-adding issues
- Adding value in research, design, development, manufacturing, marketing and sales, and business operations
- Adding value through integration
- Gains from leveraging resources

Value-Adding Issues

The term "value-adding activities" can be defined in different ways. First, management must recognize that not all activities will add quantitative value. Some value additions will be qualitative. Second, although the objective may be 100 percent value-adding work assignments, that objective may never be reached. The target is moving, and what may be considered value-adding today may not be tomorrow. As with such a concept as "zero defects," the necessary mindset is one that focuses on continuous improvement in doing things that add value. Developing that *value-adding mentality* requires that every assignment to an activity be examined for its value. Why should it be performed? What will be gained from the exercise?

The value-adding concept is that activities provide some measurable benefit in accomplishing a particular objective. Figure 8.1 illustrates the incremental nature of the value-adding process as a project flows from concept to commercialization. The designations on the horizontal axis relate not to functions but to activities. As an example, research is not the research department; it is an activity that can take place in any part of the organization. That is true of non-product-oriented projects also, except that the flow is from concept to implementation. Each activity contributes to the total value added by the effort expended. Each activity uses the business resources to provide some added benefit. That benefit may be information, resolution of a design problem, a design, or a contact with a potential customer.

A newspaper article (1991) [1] reported that a mathematician, Wu-Yu Hsiang, at the University of California, Berkeley, proved a theorem that has eluded scientists for 380 years. The problem, first investigated

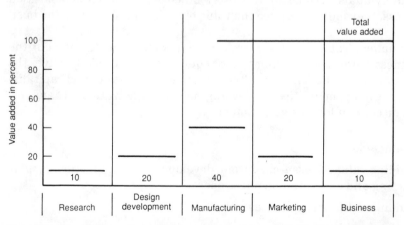

Figure 8.1 Cumulative value added by research, design and development, manufacturing, marketing, and related business activities.

by Johannes Kepler in 1611, is to prove mathematically the most efficient way to pack round objects into a square box. Kepler hypothesized that the most efficient way was to stagger the layers so that each spherical object sits in the depression formed by three spheres below it and that no denser arrangement was possible. However, Kepler was never able to prove the hypothesis. Wu-Yu Hsiang worked for 15 months, and the proof required a 150-page document that has astonished the professional mathematical community. For centuries, farmers, marketers of produce, and greengrocers have known that to get the maximum number of apples or oranges or any fruit that resembles a sphere into a square box required packing each item so that it sits in the depression formed by the three spheres below it. This was and continues to be the most efficient way of packing. Suppose that the packing of fruit in such a way had been delayed until Wu-Yu Hsiang proved the theorem in 1991. Packing fruit would have been a real dilemma. That is not in any way to minimize the contribution of Wu-Yu Hsiang to the mathematics community. It may have value in the world of mathematics or science, but to the fruit packer it was self-evident. The proof of the theorem may add value to some field of mathematics, but it does not add any value to those engaged in handling produce.

Whether an activity adds value will be a matter of judgment. When CEOs decided to introduce strategic planning (not strategic thinking or strategy), they believed that the effort added value. The question now is not agreement or disagreement. They may have been wrong, but in their view the process forced managers to look at their businesses in a more comprehensive manner. That was their judgment. The judgment of the managers was usually quite different. Although it may not have been voiced, strategic planning was looked on as a waste of time and effort: The strategic plans were seldom referred to after their approval until the next annual cycle was to be repeated. That outcome has followed the introduction of any corporate program whether related to total quality management, empowerment, or any other single issue. The following comments will provide some insight to the importance of taking advantage of the value-adding opportunities in the organizational functions. Recognizing the opportunities is usually difficult because of the behavior patterns that have been institutionalized as the right way to approach specific problems.

Value-Adding Activities in Research

Research adds value by providing answers to the unknown. Developing that new chip and finding a way to make it competitively, formulating a new material with improved properties, expanding on some basic scientific phenomenon and developing it into a usable concept, and uncovering the hidden interactions related to the sciences are examples of activities related to research. Within the list are various levels of complexity.

Although a researcher may not be able to do it right the first time, because the right way may be unknown, doing it the right way the first time, as suggested by Krogh [2], a retired vice president for research and development at 3M, is not beyond expectations. By "the right way" Krogh meant the proper use of the scientific method to yield the maximum amount of information. It involves the use of the principles of experimental design, the recording of experimental data, the observation of the process, and the interaction of the various influencing factors. The restatement by Krogh may be the more appropriate and realistic approach for the total organization. The term "decision processes" may be a substitute for the "scientific method."

If research organizations are to remain vital business functions, they must pay their way. That payment is more than just recouping their investments. The ratio of output to input must be hundreds or thousands to one, depending on the investment. Whatever the ratio, there must be considerably more output than input. Privately, some research executives complain about the lack of output of their research laboratories. There is a body of opinion that today's research investment does not correlate positively with the output, that is, the output of research laboratories is not as large as in the past even though investments have been increased significantly. One can argue pro and con about the issues and the reasons.

The AT&T Bell laboratories, prior to its dismantling, was recognized as the epitome of industrial research. The laboratories gave birth to the transistor, the laser, cosmic background radiation, and other Nobel quality efforts. It cultivated a culture in which research freedom dominated and the intent was to push the frontiers of science, primarily for exploring the unknown. Corcoran, in "Rethinking Research" [3], states that Bell is "abandoning the paradigm of research that it so zealously cultivated over the past few decades." In a sense, Bell is undergoing a cultural revolution. The culture that research was justified for the sake of research is being transformed into a culture that will find new ways to structure research, reduce duplication, and focus greater attention on application. In the transformation process, Arno Penzias, vice president of research, must attempt to balance the quality of the research effort and make sure that research pays its way. In essence, the new Bell paradigm requires adding value from the investment of research. Under the old paradigm, Bell never capitalized on its pioneering research in transistors and lasers, as well as many other contributions.

Research is but one function in the organization, and it is not an island unto itself. If its output cannot be transferred to the rest of the organization; it serves no business purpose. Research can increase its value adding by recognizing the needs of development, manufacturing, marketing, and the administrative functions. It should break down its own walls and set the example for other functions.

Research managers can increase the value added by balancing the needs of the researchers with the needs of the organization. That type of action requires a considerable amount of the right kind of pressure. Take as an example a researcher who prefers to work on something that does not meet the needs of the company's purposes and does little or nothing to help fulfill its strategy. That may be an exciting project for the individual, but if that individual is to add value from the company point of view, the work effort must be redirected.

Determining which group of projects to select and then allocating the resources raises many opportunities for adding value. As a general rule, organizations fund too many projects and none of them with sufficient resources—keeping in mind the prior descriptions of the resources. It appears at times that the guiding principle is that if you are working on a large number of projects, there are bound to be some winners. In other words, play the percentage game. That approach has neither optimized the cycle time nor provided a rational basis for continually underfunding projects. Funding fewer projects and completing them in a shorter time period adds considerable value. It helps focus the participants and creates a sense of urgency. A project load that requires frequent stops and starts not only wastes time but eventually requires additional resources to try to recover some of the lost time. That can become a never-ending spiral.

Smith and Reinertsen, in *Developing Products in Half the Time* [4], raise the issues related to overload conditions in most research departments. They classify projects as active, completed, aborted, and dormant. Their investigation revealed that between 1986 and 1990, the number of dormant projects was 4 times the number of active projects. Dormant projects drain energy from the active ones. They conclude that working on too many projects creates a dysfunctional environment. Getting a positive impact from project overload situations is a myth. As stated earlier, such management practices divert resources and extend the time to market of more viable selections.

Research must exploit the available information resources. The appropriate use of information in research provides value-adding opportunities. The process begins with the literature that is often ignored or treated as insignificant. Most researchers are known for their adherence to the principles of not invented here (NIH). That attitude continually prevents research organizations from meeting their objectives in a timely manner and decreases their value added. If a particular piece of research has already been accomplished, why delay a project by starting from ground zero or take the time to try to modify the concept and in the process complicate the research effort?

Creativity and invention are necessary outputs of research. Their results add value to the organization. But eventually the results of that creativity must be supplemented by a large infusion of other resources. Demonstration of feasibility has its limits. It is at this demonstration point that organizations begin to invest a significant amount of their resources in attempting to commercialize the idea. Researchers have an opportunity to add value by fully disclosing the known limitations and potential problems. Romancing a new idea or concept provides opportunities for exploring the full potential of the concept, but at some point an injection of realism is necessary. At that time, research must neither overstate nor understate the limitations. Realistically appraising the state of the art of the new discovery adds value.

Adding Value in Design and Development

The designers and developers of new products add value by the quality of their designs as related to manufacturability and meeting customer requirements. Too often, organizations lose sight of the level of creativity required in the design function. Some examples will show how design and development can add value that increases with each production batch.

It is one thing to take a current set of engineering drawings, make some relatively minor changes, and call the result the design for a new product. It is quite different when the designers begin the process with blank sheets of paper before them and depend on their past experiences and level of creativity to give birth to something new. The ability to add value during the design process is unlimited. Over the years, organizations have attempted to do value engineering to determine the quality of the design. But all that is after the fact. The design already exists, and any value engineering will require rework. The objective must be to incorporate the value-adding concepts in the design during the design process, and not after the completion of the design or test of a developed prototype.

Product designs normally include a multiplicity of technologies. Achieving some end involves either a mechanical motion or transformation of some type of energy. At the very least, mechanics, electronics, and selected frequencies of energy from the electromagnetic spectrum interact in some form. Making the trade-offs provides opportunities for adding value. In the design of an automatic machine, the drive system can be cammed from a single drive source or individual synchronized drive motors can be controlled through a complex electronic control system. Each approach has advantages and disadvantages. A similar statement applies to the decision to use the principles of continuous motion or intermittent motion. Which approach is used depends on the specific application. The choice makes a significant difference in value added. Bottle-filling systems operate by using continuous motion; that is, the bottles are filled, capped, and labeled while they are in motion. The intermittent motion approach would bring a series of bottles to a fill station, go through the fill cycle, and then move to a capping station and possibly to a labeling station. The approach taken by the designers will determine the value-adding opportunities of their design.

Design and development add various amounts of value depending on the complexity that the design creates in the manufacturing process. Designs that require excessive or specialized tooling must be carefully evaluated. Specialized or high-tolerance tooling may require a great deal of experimentation that can extend the desired cycle time. Design for manufacturing is the art of compromise between the elegance and the requirements of the design. Without any question, design elegance is often required in precision products or equipment, but adding unnecessary elegance reduces the value-added benefit of the design.

Conceptual design is one element of the up-front work that must be performed with a high degree of discipline if the follow-up work is to flow smoothly through the system. It is at this stage that the concept must be fully described, the critical design factors and the unresolved problems enumerated, and general agreement among the major product functions reached. Until the conceptual approach is clearly delineated, it is almost impossible to begin any major design activity. That approach is foreign to U.S. organizations. People who authorized the project want to see something happening, and they do not believe anything is really happening unless they can see metal being cut or purchase orders issued. They talk about the "fast-track approach," but the scale-up of the product in manufacturing seldom reaches its goals. Engineering and manufacturing changes waste resources, increase cost, and increase cycle time.

Adding Value in Manufacturing

Manufacturing operations provide many opportunities for gaining significant cost benefits by focusing attention on value-adding activities. Manufacturing equipment has become more complex in the past decade, but organizations continue to use the same methods for resolving the array of problems normally associated with bringing equipment on-stream and managing the uptime. No manufacturing manager will discount the cost or the problems associated with the startup of any new process or product. Even though the new technologies are more complex, they continue to use the same practices that were used when the expectations of technology and product performance were relatively simple.

Consider as an example a complex automated assembly process that includes not only the assembly of the parts but the final testing of the product and the packaging. The expected output may be anywhere from 50 to 1000 pieces per minute from either an intermittent or a continuous type of machine. It is not uncommon to see a group of people, usually including the plant manager, the head of maintenance, several maintenance people, and even some engineers, huddled over some portion of a machine attempting to determine why the system continues to shut down intermittently or why defective product is coming from the end of an assembly line. Those huddles seem to be a daily occurrence. After an hour or two, someone makes some adjustments and the machine continues to chug along for the rest of the shift. The same situation occurs on the second and third shifts, and the process is repeated day after day. The problems are usually different, but the same group meets and really fiddles with the machine rather than determine the problem. This non-value-adding activity may require working overtime to meet the production schedule.

Why are such scenarios so prevalent? Because no one in the plant or machine design area recognizes that the machines are operating too fast for the human eye to perceive not only the action but the interaction between component parts. Complex assembly machines involve timing such that the eye cannot perceive the exact placement of parts, component part defects, vibration problems, and so forth. There are high-speed motion analyzers that can be used to pinpoint such problems; they record actions that the human eye cannot perceive. The images can be replayed frame by frame on a video terminal or a hard copy can be made if necessary. Such motion analyzers provide information that allows engineers and maintenance people to arrive at some conclusions. Resolution of the problem is not left to such approaches as "Well, let's try this or that" without any evidence that what is tried is even applicable.

What is the value added by a group of people huddled over the machine attempting to resolve the problem? Zero. The people do not have the tools necessary to determine just why the problem exists. Management may have approved $10 million or more for the machine, but the $50,000 to $75,000 for the test and measuring instrumentation was not deemed necessary by manufacturing. The loss in value-added activity by the personnel involved with the problem is only a small part of the cost. The losses due to machine downtime, waste and scrap, and working overtime is far greater. The same scenario can be found in the operation of a complex machine or a simple stand-alone piece of production equipment such as a lathe or mill.

Continuing with the same scenario, it is well known that automatic assembly machines require an input of parts that meet specifications of tolerances provided in the assembly machine. Parts that do not meet machine specifications create stoppages that increase the downtime. In discussing the requirements for an automatic assembly machine, the manufacturing contingents became adamant about the a requirement that the machine could stop only once in every 40-minute period. An inspection of a batch of only 1 piece part out of 12 in the assembly indicated that the machine would probably stop every 3 seconds. The example demonstrates that in such a situation, unless the piece part tolerances can be improved, little if any value will be added by automatic assembly equipment.

Inventory management provides an example of an often-ignored parameter in manufacturing and its subsequent impact on performance. Many organizations talk about just in time (JIT) as a means of controlling inventories at all manufacturing levels, but the results have not been encouraging except in isolated cases. A study by Rabin and McGrath [5], reported in "Eleventh Annual Study: Assessing High-Tech Inventory Management," shows that, in 1988, inventory in the high-tech industries remained essentially constant. The impact of inventory management can be demonstrated by reviewing the results achieved by Digital Equipment Corporation during the period 1984–1988. Inventory turns increased from two to just under four and provided a $1.5 billion benefit to cash flow. How Digital's management could live with an inventory turnover of approximately two in 1984 raises some serious issues about the competence of senior management. A follow-up study by Rabin and McGrath [6] for 1989 showed an improvement in inventory turns from 3.9 in 1988 to 4.1 in 1989, an increase of 4.6 percent. Rabin and McGrath state that:

> If there has been a downside at all to the widespread acceptance of Just-In-Time (JIT) concepts in American companies, it is that it has, in many cases, prolonged the view that the manufacturing function can independently optimize its own operations. Although any manufacturing strategist knows this to be untrue, the manufacturing function has historically "bootstrapped" itself through improvement program after improvement program.

If JIT is going to continue to provide benefits, the concept must be implemented throughout the organization. It is not solely a manufacturing program. All the functions that involve the movement of resources must become involved to achieve the optimal benefits.

Adding Value in Marketing and Sales

The final measure of how marketing and sales add value can be quantified. There are also qualitative issues that become quantified in the process. Marketing and sales involves bringing in the orders, shipping the product, and making sure that the customer is satisfied. Customer satisfaction requires more than not hearing from the customer about any complaints. It demands contact with the customer to determine the level of satisfaction. In many cases, marketing and sales are the final determinants of a new product's level of success, but marketing and sales must take a systems approach to add value.

In Cooper's research into the reasons why products fail in the marketplace, 13 key activities were identified (Chap. 4). Five were related to marketing and sales:

- Lack of market assessment
- No detailed market study or market research
- No customer tests of the product
- No test market or trial sell
- No formal market launch

The mortality rate of new-product introductions is well known. It is difficult to argue with the five reasons for product failure attributed only to marketing and sales. Marketing and sales add value by performing their activities diligently and also from the requisite knowledge base. They are self-evident requirements for success.

Product superiority and low cost are not guarantees of success. Inferior products will not survive in the marketplace, but superior products have failed for many different reasons. Market dominance is short-lived in these days of international competition. At one time, Kodak and IBM dominated their markets and called all the shots. That is no longer true, as demonstrated by the results of both companies in recent years. When an organization establishes the price for a product and dominates the market, what is often translated as excellence is really ineffective and inefficient operation that is clouded by the total revenue stream. Sales revenue and profits continue to grow until the organization no longer holds all the cards.

Quality, performance, and service are expected. Most reliable organizations will provide various levels of each, but they are a must. The needs of consumer and industrial products may differ considerably. The auto industry disregarded its customers for many years, and it is paying the price as it attempts to restore customer confidence. However, when industrial products are purchased according to a performance specification, the situation is quite different. Quality, performance, and service are (or at least should be) a must. The issue was raised previously about the differences between order takers and salespersons. The primary difference is in the way in which the customer is treated. Customer relations do not end with the order. How a customer is treated after the order may be far more important if business growth depends upon repeat orders. Perhaps one of the most difficult activities in marketing is to review the customer base periodically. Sales may be growing, but who are the customers? Replacing one customer with another, regardless of the effort, does not add value. It may add sales, but the question that must be asked is why so many customers have switched to competing products.

Adding Value at the Business Operations Level

Countless activities are requested by executives, managers, and other persons in supervisory positions. The requestors would argue that their requests may be exploratory and therefore essential. The activities that result in new usable information or action are without doubt essential and add value. The requests that should concern everyone are those that add no value and are spur-of-the-moment. Strategic planning has been suggested as a program that did not add value commensurate with its use of resources. Excessive planning of any type falls in the category of not adding value.

It is easy to focus attention on the activities that do not add value. In reality it is just a way of thinking. One can ask what is right or what is wrong. It is only a matter of whether the glass is half full or half empty. At the business level it is only necessary to ask a simple question: What was the net effect or impact of my activities? What did I contribute to the organization? Managers must ask themselves: How was business performance affected by the use of resources in my department? What did the department contribute to the business, not the department? Such simple questions require a great deal of honesty. Can a person admit that his or her effort had little or no benefit on the success of the organization? Can a manager admit to himself or herself that the department did not add value to the organization commensurate with the resources expended? Can managers admit to their senior executives that the consultants who received half a million dollars for evaluating and making recommendations did nothing more than massage the inputs from their own people? How can a manager admit to colleagues and senior management that little if anything was accomplished by the consultant?

What kinds of activities add value at the business level? It is impossible to list all such activities, but some suggestions may help stimulate the thinking about the kinds of activities that add value. All the examples relate to anyone in a managerial position.

Business reviews

Consider the business reviews at different locations, including international, that are either dog-and-pony shows or opportunities to clarify major issues, transfer information, and provide guidance. They either add value or can be dismissed as junkets. They add value when they are spontaneous and the reviewers really come to see it as it is. They make their contribution either by applying an understanding of the major concerns or by using their past experience to coach or provide a path to the resolution of the concerns. They waste resources when they are preplanned and everyone knows what the review team is concerned about—the group making the presentation knows which buttons to push to divert attention from the real issues. So the canned presentation takes up most of the time, and little time is reserved to involve the reviewers in the solution to the problem.

Approval or authorization process

The approval process can either add value or waste the business unit's resources. In any major corporation, the approval process for major investments of any type may require signatures on several layers of management. The approval process can add value if those involved in it go beyond considering the return on the investment figures. The approval process can either be a rubber stamp on some previously quantified information or an understanding of what the investment means to the organization. Predetermined financial hurdle rates generate lost time. The approval process usually involves some level of controversy (or at least it should), because someone must ask the difficult questions. Those questions must be raised and answered. The risks should not be a surprise at some time in the future.

Allocating resources

Most organizations seldom if ever emphasize the need to allocate resources in a way that optimizes the value added. The assumption is that the activities are adding value; otherwise, the organization would not be meeting its financial targets. That argument loses validity anytime the objective is the ideal or theoretical potential rather than a competitor's performance or last year's corresponding quarter. Focusing an organization's attention on value-adding activities is a significant value-adding activity for the CEO and top management: Action must begin at the top, and any discussion must be reinforced by that action. Levels of performance and productivity would not be a current concern if in the past a major effort was directed at activities that add value.

Technology

Technology is a specific area that requires a focus on value-adding concepts. Investing in technology is an expensive hobby. The word "hobby" is used because organizations think in terms of technology residing in boxes labeled "research," "development," or "manufacturing." It is difficult to find an organization that has a business unit technology plan that unites the technologies of the three functions. Gaynor [7] presented a description of business unit technology plans as a means of integrating the technology functions into a business rather than a functional perspective. Unifying the technology plan in such a way as to show the relations among technologies adds value by focusing the organization on its purposes, objectives, and strategies. It should eliminate the need for the participants in the technology areas to waste time and effort vacillating and waiting to see from which direction the management wind will be blowing.

Information transfer

Figure 5.4 illustrates the deviations that normally occur between the ideal and the actual transfer of information as related to new product introduction. The extension of product cycle time is directly related to how an organization adds value through allocation of its resources. The ideal curve is not a dream or an artifact from fantasy land. If organizations operate according to the pattern shown in Fig. 5.5, whereby the primary new-product functional groups recognize their responsibility for their contribution in the early stages of a project and continue until their services are no longer required, the ideal curve becomes a probable target for accomplishment. Pushing to meet that ideal curve means adding the required value at each step in the concept to commercialization process. The curve, with the discontinuities, shows what occurs when the value-adding activities of the various functions are not synchronized.

Information technology has been praised and condemned with equal vigor. As the computer companies try to outdo one another with the introduction of the fastest with the mostest, organizations fight a paper overload never before seen in U.S. industry. There is a simple answer for the dilemma: Executives were sold a bill of goods. More functions and reports were added without questioning the value of the reports. The added value was not considered. Organizations continue to struggle with improving white-collar productivity, that is, adding value, by using management information systems only to find themselves making greater investments that result in lower productivity. Davis, in "Information Technology and White Collar Productivity" [8], says, "Despite the hype that has accompanied the introduction of the latest information technology, white-collar productivity has not shown appreciable improvement."

There is no one single reason for the complaints leveled at management information systems, but there are possibly two significant reasons that are closely related: (1) Large-scale computer systems have failed to meet the specifications, cost, and schedules. (2) Organizations continue to focus on data rather than information or knowledge.

Davis provides an example of a situation that is generally the rule rather than the exception. In 1982, Allstate Insurance, a Sears' subsidiary, decided to build the most sophisticated computer system in the insurance industry. Many of Allstate's office operations were to be automated. One of the objectives was to shorten the time needed to introduce new policies. Allstate hired Electronic Data Systems (EDS), a systems integration company, to develop the software and install it on the firm's hardware. The target date for completion was 1987, at a cost of $8 million. By November 1988, the project was $15 million over budget and still not completed. Allstate terminated its contract with EDS and set a 1993 completion date.

Management information systems continue to generate more paper. Perhaps much of that is due to the fact that organizations are attempting to eliminate the judgment factor from managing. Computers are being used when a pencil and piece of paper would suffice. An immediate telephone response is substituted for the monthly report. The Frito-Lay situation described in Chap. 2, in which a newly installed data management system for collecting field sales information was credited with warning the organization about the encroachment of a competitive product, is a good example of a misuse of information systems. In that case the salespeople must be either inept or totally untrained. Part of a well-trained salesperson's responsibility is to keep his or her antenna tuned to the customer forum. What should have been reported immediately took several weeks. The organization was focusing on formalized data reporting rather than immediate feedback from the marketplace. That the DDS took several weeks now instead of the 6 weeks then to collect the same type of information hardly justifies the computer system. The problem is not lack of information; it is inadequately trained salespeople.

Many theoretical approaches to management of information have been taken. Each is presented as the ultimate answer to management performance. Rockart and Short, in "IT in the 1990's: Managing Organizational Independence" [9], argue that the role of information technology (IT) is "allowing firms to manage organizational interdependence." Current literature and, even more important, management practice considers information management from four perspectives:

- Internal structure: how it affects roles, power, and hierarchy
- Organization of teams: changing work groups using information technology
- Disintegration of borders: ease of electronic communication among firms, suppliers, and customers
- Technical: systems integration of business processes across functional, product, and geographic lines

Rockart and Short add a fifth perspective: managing interdependence as a means for responding to competitive pressures. They describe *effective interdependence* as the firm's ability to achieve concurrence of effort along multiple dimensions of the organization. In essence, they support this author's conclusions that firms must adopt a systems approach to managing, integrate not just information but also management, organization, and technology, and find ways to arrive at a more nearly optimal point on the continuum from excessive differentiation to restrictive integration. That balance must be regained before organizations can hope to approach their optimal performance.

People: The origin of added value

The primary barrier to adding value is people: parochialism, the NIH attitude, not knowing the contribution that others can make, and lack of breadth of knowledge of how the system functions. Parochialism may be the greatest barrier: It locks out everything from the outside. The doors to alternative actions are closed. Isolation seems to dominate the culture. Keeping activities with a certain person or in a specific department when those activities could be performed more effectively by others or in another department not only prevents the optimal uses of people but may lead to early obsolescence.

Emphasis on people must be the primary concern. The intellectual resources reside in people and their ability to call to mind the past experiences of the organization. People generate and use information: How they generate and use it determines business performance. People, not statements from the executive offices, determine culture. People determine the relations between the customer and the organization. People are the sources and implementers of technology and then the users of the products or services that stem from that technology. By their level of performance, people determine the financial strength of an organization. People design plant and equipment and then use those resources to provide the products or services to the marketplace. People work in facilities and use the facilities to make their specific contribution to the business. People use time to pursue their activities at various levels of effectiveness and efficiency in meeting business purposes and objectives; being in motion does not necessarily translate into making progress, and dreaming does not necessarily imply a waste of time. In the final analysis, people determine the level of success in any organization. Managing the activities of the people resources must therefore take precedence. People are the source from which all actions arise. However, a distinction must be made between managing people and managing their activities. Activities can be managed. It is doubtful that people can be managed and at the same time be expected to use their initiative, creativity, and intelligence—at least not for very long without some threat or fear.

Adding Value Through Integration

Additional value can be added if the activities of the functional and multidisciplinary groups can be supplemented by an interdisciplinary approach. If, as an example, research, development, and manufacturing could begin functioning as one, rather than independent functions, value of their activities would be increased. Integration within functions and between functions internally, coupled with integration with external supporting operations, will add value. But what should be integrated? The answer is simple: Everything. Integration cannot become another case of single-issue management. Integrating strategies is insufficient. Integrating purposes, objectives, and strategies is insufficient. Integrating research with manufacturing and marketing is insufficient. Integration must be viewed in comprehensive terms.

Organizational segmentation and integration lie at opposite ends of the spectrum. Integrating the activities of the firm presents managers with complex decisions. Although most organizations function at the differentiated or segmented end of the spectrum, moving to a position of total integration, which is equivalent to consolidating all related activities, would limit the effectiveness of the organization and create a set of different problems. Senior management is responsible for finding the appropriate position on the spectrum for a particular set of circumstances.

Whiston, in a series of four articles published in *Technovation,* considers some of the broader issues of integration as related to performance. Part 1 provides an introduction to and overview of integrating the technologies of a manufacturing organization [10]. Part 2 presents a managerial and organizational perspective [11]. Part 3 focuses on the conditions for managerial and organizational integration within an organic enterprise [12]. Part 4 considers the remaining policy issues and organizational questions related to education, behavioral practices, and the need for a critical mass and illustrates the micro-dependencies influencing integration [13].

Whiston cites Voss [14], who subdivided integration into a hierarchy consisting of five facets: strategy, material flow, technology, information, and organization. Voss and Whiston focus attention on integration within manufacturing; Whiston added managerial integration to the list. Although both authors consider those facets in a hierarchy, it is necessary to recognize that all such categorizations contain elements of artificiality. It is not possible to develop a theory that would apply in all situations. Success in a business requires more than integration or the contribution of some specific act of integration that may be difficult to identify. How those aspects are arranged in a hierarchy is not important. What is important is that all be brought into the process; none can be ignored.

As organizations attempt to integrate all of the resources and all of the elements of the infrastructure more effectively, managers must recognize the need for new competencies. Much of the failure resulting from investments in new technologies in manufacturing came about as a result of insufficient competence. Even such relatively minor changes as the integration of equipment requires new competencies and new approaches. If integration involves new processes and advanced manufacturing technologies, the benefits of integration will not be achieved until the participants are intellectually prepared for dealing with problems under different operating conditions. Changes in organizational structure, changes in the role and responsibilities of every participant, and acquiring new knowledge, skills, and competence will be required to effectively interact with the resources and the infrastructure. Success in integration is limited only by cognition and the effectiveness in implementation. A quotation from Whiston summarizes the conditions required for integration:

Integration demands an intellectual perspective, with regard to process and product development, which not only links more closely the manufacturing strategy with corporate strategy, and links research and development policy with marketing and manufacturing functions, but also continually questions the opportunity costs of increasing skill differentiation and functional specialization. Specialization and professionalism are very necessary. However, the interlinking of cross-functional integration of such skills remains an increasing managerial and organizational challenge.

Leveraging the Business Unit Resources

Leveraging the business resources provides an additional means of adding value. It involves combining the resources and using them in such a way as to gain the most advantage from the combined effort. Resources can often be substituted, primarily when consideration is given to developing internal resources when external resources may already be available. People can be substituted at times for new major capital investments, and vice versa. Outside services of all types can be provided on a contractual basis. Peters [15] talks about the vertical deintegration by which, at some time in the future, multibillion dollar corporations may contract for all of their services. To what extent an organization should use external resources depends totally upon the financial and long-term business considerations. It is not an all-or-nothing approach. Different circumstances require totally different solutions. The simplistic answer does not work.

There are no figures to demonstrate how organizations use their resources. Waste and scrap are measured in the factory but seldom in other business activities. Some examples of how to leverage resources and the consequences of not leveraging may help focus management's attention.

People resources

Leveraging the knowledge base of an organization is a critical factor in adding value. The story is told about the exceedingly competent optical physicist who was developing a measuring instrument. His specialty was optics, but the system required in-depth knowledge of mechanics, flow systems, and the design of solid-state circuitry. He refused assistance from specialists in those other fields and, unfortunately, his management permitted the situation to continue. The physicist even went to the point of attempting to design his own solid-state devices, a field he was not familiar with. After 4 years, the project was terminated. If the right complement of talent had been assigned, the project probably could have been completed in less than 2 years. The technology was available; the physicist had only part of the answer. Managers who feel that this is an extraordinary case should seriously look at their own list of projects and the assigned people. The same kind of situation is found in all other organizational functions. Making use of all the knowledge sources, regardless of origin or location, prevents reinvention of the same wheel and leverages the people resources.

The *Harvard Business Review* [16] reissued an article written by Schaffer and published in 1974 titled "Demand Better Results—and Get Them." In it Schaffer raised the issue of demanding better results. In a retrospective commentary on the 1974 article, Schaffer makes several points that are of greater importance today than in the past:

> Ironically the *thinkers* who have invented the latest organizational effectiveness strategies unwittingly provide new busywork escapes. By putting so much emphasis on processes and techniques they have slighted the importance of results. Thousands of employees are trained in seven-step problem solving and statistical quality control; thousands of managers are *empowered*; and thousands of creative reward and communications systems are in place. In the absence of compelling requirements for measurable improvement, however, little improvement occurs.

Schaffer goes on to stress that setting high standards of performance does not conflict with empowering people because empowerment comes as people rise to the challenge of new demands. Challenges in industry do not occur as they do for three hours in a football game. Meeting business challenges is an everyday event; it is what business is all about. People who are challenged in their work activities soon find themselves achieving more than they expected.

Intellectual resources

People, as well as the business unit archives, are a repository for their intellectual resources. 3M's Post-it notes [17] is an excellent example. The construct of the adhesive was an intellectual resource put on the shelf because of no known application at the time—the researchers were not looking for an adhesive with those particular characteristics. However, observation and recognition of potential value at some future time are critical characteristics of the researcher. Organizations must resurrect their intellectual resources. The scenarios of development departments going ahead without understanding the past continue to be replayed every day. The intellectual resources of the marketing group are of equal importance. At the same time, manufacturing has a base of intellectual resources that can be tapped to ensure that products can be manufactured at the required cost. In a systems approach, the intellectual resources of all those functions, when brought together and used appropriately, can leverage the advantage of any single resource.

Information resources

More and more is written about information as a means of achieving competitive advantage. There is no doubt that information adds value—not data, but information converted into knowledge. Whether information by itself will achieve competitive advantage is certainly questionable. Without doubt, information is important, but it does not entirely reside in the corporation's database. Information that is generic to the business can be leveraged and impact all the factors that add value. As an example, information about the field results of a new product can provide an input to research, development, manufacturing, and anyone involved in the new product process. It is not the sales or financial information that is vital to the participants; it is the performance of the product as witnessed and reported by customers. That, if known and used appropriately, is the feedback necessary to determine future improvements or the introduction of totally new products. The information must be timely if it is to leverage resources and add value. The feedback must be immediate. It may not even enter the management information system. People must be educated in the need for transmitting relevant information. Granted that relevance is a matter of judgment, employees must be informed about what may or may not be relevant.

Cultural resources

Culture became the managerial quick-fix after the publication of Kanter's *The Change Masters* [18]. Popularization of a concept tends to reach the absurd; her work deserved better and closer attention. As has been stated, culture must somehow change behavior. It must be more than just the recitation of the goodness feeling. The downturn in the economy since 1989 has raised havoc with the cultures of prominent organizations. When organizations controlled the markets and the technology, many accommodations were possible. Performance expectations were modest, and companies developed cultures that eliminated personal risk and excused all levels of nonperformance. Once on the payroll, the employee was there for life. As competition increased, the same companies found themselves with the task of making significant reductions in personnel. The company was no longer family. Those cultures will now have to change. The behavior pattern regarding expectations will mean being pushed to perform more effectively by raising the performance expectations to where they should have been. How does an organization leverage the culture of the past into a new culture that requires continual emphasis on high levels of innovation and performance? One approach for certain will not work: executives placing the responsibility for not meeting performance expectations on lower-level managers, professionals, and operating personnel. Executives created that culture either naively or by neglecting the fundamentals. Reorienting the culture must begin at the top, flow through the organization, go back up to the top, and so on through many cycles. The point is to understand and really know the key elements of that past culture. Not all of it needs scrapping. Most of the employee-supportive cultures were excellent. The difficulty was that they went too far: they gave and gave, without any expectations in improved performance. In many cases, the employees themselves recognized that they were not being challenged by management. Senior management did not take advantage of the culture by raising expectations; in the process, it failed to add value through the policies and practices that it espoused.

Customers as a resource

The input from customers can be leveraged if an organization is sensitive to customer needs. The argument over whether businesses provide customer needs or wants will continue, but it is academic. The fact remains that people buy to satisfy either wants or needs. Few people need video cameras; the cameras are generally a want that is satisfied. Sony and its introduction of the Walkman provides an example of leveraging customer input to the point at which it became a Sony product strategy: meeting the needs of specific market sectors with specialized products using common technologies but in different manifestations. Sony leveraged the information it received from its customers. That leveraging added value by providing the input for a continuous flow of products. Customers can also provide ways to enter new or allied markets. All that is needed is the ability to see beyond the immediate. Observation provides new opportunities. How can new technologies and needs be packaged differently to provide the customer with a real benefit? Customers are a resource that can be leveraged to provide an added value.

Technology resources

Every organization operates with a number of core technologies. An organization like Kodak's photographic operations is involved in coating technology, emulsion technology, optics, and a series of specialized manufacturing technologies such as molding and assembly. Most of Kodak's photographic products use the same coating technologies. A new process is not required every time a new or improved product is introduced. The converting and assembly operations may require totally new facilities, but Kodak leverages its coating technology basically through continuous minor improvements. That is true of companies like 3M, Fuji, Agfa, and Konica also. Most batch chemical processes and some continuous processes can also be leveraged to produce a wide variety of raw materials.

But some new thinking is required to learn how organizations can leverage new technologies more effectively, especially the soft technologies in the administrative functions. Performance appraisals were singled out as a form of human resource technology. How can the total effort that requires countless hours be restructured to provide a greater benefit? There is no doubt the appraisals are important, but a current cost-benefit study would not yield a positive result from either a qualitative or a quantitative point of view.

Financial resources

The question to be asked is how an organization optimizes the use of its financial resources. Leveraging is considered not from the standpoint of investing excess funds for short-term benefit and so on, but from the intention to invest in the resources for sustainable growth and profitability. As an example, consider a major investment in new plant and equipment in the multimillion dollar range. Evaluating an investment of $20 million or more requires knowledge of the business as well as a clear understanding of the availability of adequate resources at the time of the investment decision. First, it is important not to confuse financial projections with facts. Finance people educated in company products, processes, markets, and so on can leverage their impact on adding value. The extent to which they leverage their input depends on when they join the process. If they become partners in the early stages and are a part of the team developing the project, their continued input and participation add value when needed and not after all the preapproval work has been completed. They have the figures from past performance and can educate the technical and marketing people about the facts of life—the business reality. By being part of the team, they are not sitting in judgment; they are providing an input to determine if the investment is viable and within the financial resources of the organization.

Plant and equipment resources

Leveraging plant and equipment provides different types of opportunities for adding value: reduced product cost through greater utilization, reduced space requirements, more flexible equipment, and so forth. Plant and equipment can be leveraged through the use of advanced manufacturing technologies, which must be implemented from a systems perspective. As an example, computer-integrated manufacturing has failed to meet expectations in most cases. The computer junkies managed to get control, and they often computerized inefficient and unproductive manufacturing operations that provided little if any real additional benefit. Initially, the organization's equipment could not meet the production requirements; after computerization, the organization was the proud owner of a computerized inefficient and unproductive manufacturing operation.

Plant and equipment can be leveraged by greater utilization. In the future, this will require new approaches to the design of plant and equipment. Idle plant and equipment adds cost. Making a major investment in plant and equipment that will operate 8 h/day for 5 days/week requires careful attention. An example of unused capacity is Steve Jobs' NeXT workstation plant, which is probably the most automated manufacturing system in the computer industry. According to *Fortune* [19], the plant could produce $1 billion worth of workstations per year with no more than 100 workers. The 1990 production was about $100 million in hardware. That represents a lot of idle capacity.

There are two aspects to leveraging plant and equipment. One is obtaining the maximum use of an organization's own investment; the other is making use of other organization's productive capacity. In the first situation, if production capacity is limited, using outside contractors provides a possible alternative. If excess production capability exists, firms can consider entering new businesses in which the current capacity would be better utilized. In the second case, no investment is necessary—or certainly nothing more than specialized tooling. Both approaches provide value-adding benefits to the organization.

Facilities resources

Two examples will demonstrate how facilities can be leveraged and in the process add value. The first is the purchase of computers or terminals. If a running-time meter were installed on some computers, not many hours would be clocked in any one year. An argument could be made that running time is not a measure of importance. On the contrary, it is. The computer was an investment of the financial and people resources, and it must have some type of payback. Although office managers argue that a computer terminal is a requirement for everybody, that does not necessarily mean that every person must have his or her personal computer. The only need may be access to a computer or a terminal. Workstations can often be arranged in such a way as to reduce the number of computers or terminals by at least a third and often by as much as 50 percent. That means one-half the investment and one-half the maintenance cost without any lack of accessibility.

Working facilities play an important role in determining attitude, which in turn affects the value added by each employee. Although there is no need for luxury, facilities must be adequate and comfortable. Facilities alone, regardless of their level of opulence, provide no real benefit. The most up-to-date laboratory facilities do not necessarily stimulate scientists and engineers to heights of creativity. As a matter of fact, they can be a deterrent. Creative people must try things. They do not just sit and create by thinking. They may be more creative and more comfortable in an environment that might be abhorred by others. Working facilities must in some way meet the needs of the individual. Such facilities provide the stimulus for the file clerk as well as the scientist. The difficulty lies in providing the right environment for the individual. Working facilities leverage people performance.

Time as a resource

Time is a resource that can be leveraged once organizations stop watching the clock. That is not a contradiction. Most organizations work normal hours, even those that have installed flextime. Take as an example the work of a designer: working 5 min over the normal working hours can often save an hour the next morning. If that situation occurs—and it is very common—working the 5 min adds value. That applies to every function and every activity. Every time the work cycle is broken, time is taken to go back and determine what mental processes were in place when the work stopped.

Using the intellectual or information resources of others leverages an individual's time. Using the manufacturing facilities of others leverages the organization's time. Using available technologies of others leverages time. Education and the elimination of interfunctional barriers also leverage time. All those actions add value.

Summary

Emphasizing value-adding activities would appear to be one of the more logical operational tenets of good management practice. The difficulty that arises is determining what activities add value. Will managers, before making a request of any type, think whether the result of the request will add value? Optimizing cycle time requires that all requests, regardless of origin, be questioned intelligently and evaluated against the purposes, objectives, and strategies of the business unit. That evaluation requires a response to several questions:

- What is the activity? Has it been described adequately?

- Why is the activity important or not important? Why do it?
- How does the activity provide a benefit? Does it add value?
- Who is going to do it and who is going to use the output?
- When will the activity be completed and when will the result be available?
- Where will the results be implemented?

Value is also added by integration within and between the functional groups. Value can also be added by integration with all the related external groups. Integrating multidisciplinary and multifunctional activities into an interdisciplinary working operation remains the challenge of management. There are no simple answers, but reason indicates that lost time, currently accepted by managers as a given, detracts from meeting the requirements of optimizing cycle time.

If managers can focus attention on using the resources and the infrastructure to add value and then integrate the independent functional perspectives into a business-oriented perspective, they can then attempt to add more value through leveraging. Leveraging involves finding new ways for optimizing the output of available resources. It involves making trade-offs between resources. It involves a process and a new attitude toward the business from a system's perspective.

References

1. "A 380-Year Puzzle," Reprint from *The Los Angeles Times* in the *Minneapolis Star Tribune,* July 31, 1991.
2. L. C. Krogh, "Measuring and Improving Laboratory Productivity/Quality," *Research Management* now *Research•Technology Management,* November-December 1987, pp. 22–24.
3. E. Corcoran, "Rethinking Research," *Scientific American,* December 1991, pp. 136–139.
4. P. G. Smith and D. G. Reinertsen, *Developing Products in Half the Time,* Van Nostrand Reinhold, New York, 1991, pp. 94–99.
5. Pittiglio, Rabin, Todd, and McGrath, "Eleventh Annual Study: Assessing High-Tech Inventory Management," *Production and Inventory Management Review with APICS News,* July 1989, pp. 42–45.
6. Pittiglio, Rabin, Todd, and McGrath, "Twelfth Annual Study: Assessing High-Tech Inventory Management," *Production and Inventory Management Review with APICS News,* July 1990, pp. 34–36.
7. G. H. Gaynor, *Achieving the Competitive Edge through Integrated Technology Management,* McGraw-Hill, New York, 1991, pp. 139–161.
8. T. R. V. Davis , "Information Technology and White Collar Productivity," *Academy of Management Executive,* vol. 5, no. 1, 1991, pp. 55–67.
9. J. F. Rockart and J. E. Short, "IT in the 1990's: Managing Organizational Interdependence," *Sloan Manage. Rev.,* vol. 30, no. 2, Winter 1989, pp. 7–17.
10. T. G. Whiston, "Managerial and Organizational Integration Needs Arising out of Technical Change and U.K. Commercial Structures. Part I: Introduction and Overview," *Technovation,* vol. 9, 1989, pp. 577–605.

11. T. G. Whiston, "Managerial and Organizational Integration Needs Arising out of Technical Change and U.K. Commercial Structures. Part II: A Managerial and Organizational Perspective," *Technovation,* vol. 10, no. 1, 1990, pp. 47–58.
12. T. G. Whiston, "Managerial and Organizational Integration Needs Arising out of Technical Change and U.K. Commercial Structures. Part III: Managerial and Organizational Integration within the Organic Enterprise," *Technovation,* vol. 10, no. 2, 1990, pp. 95–118.
13. T. G. Whiston, "Managerial and Organizational Integration Needs Arising out of Technical Change and U.K. Commercial Structures. Part IV: Remaining Policy Issues and Organizational Questions," *Technovation,* vol. 10, no. 3, 1990, pp. 143–161.
14. C. Voss, "The Managerial Challenges of Integrated Manufacturing," Paper read at *U.K. OMA Annual Conference,* Dunblane, Jan. 5–6, 1989.
15. T. Peters, *Thriving on Chaos,* Harper & Row, Publishers, New York, 1987, pp. 146, 475.
16. R. H. Schaffer, "Demand Better Results—And Get Them," *Harvard Bus. Rev.,* March-April 1991, pp. 142–149.
17. P. R. Nayak and J. K. Ketteringham, *Breakthroughs!,* Rawson Associates, New York, 1986, pp. 50–73.
18. R. M. Kanter, *The Change Masters,* Simon and Schuster, New York, 1983.
19. M. Alpert, *Fortune,* February 26, 1990, pp. 75–81.

Chapter

9

Barriers to Exploiting Cycle Time

The primary barrier to exploiting cycle time is management: first-line managers, middle managers, and top management including the CEOs. Too many executives and managers continue to be occupied with minutiae. Perhaps that preoccupation has been fostered by the business media and the schools of business. Journalists often know little about the complexities of managing, and they are infatuated with each new management gimmick without really understanding the underlying principles. Yesterday's heroes become today's front-page drifters. There is no shortage of articles on Japan, the Baldridge Award, delegating, charm schools for managers and executives, culture, leadership, innovation, entrepreneurship, empowerment, and even selecting the CEO with the lowest cholesterol level. All those good things are supposed to make industry competitive. Advice also comes from academic research, which has fostered the touchy-feely approach to managing: Make sure that employees are happy. That state of euphoria is accomplished by providing many benefits and accepting moderate to barely acceptable performance.

Many of the barriers have been mentioned: lack of management focus and direction, narrowly defined strategies instead of total business strategies, insufficient knowledge of how the organization functions, inadequate human resource practices, disabling interfaces, not managing according to the three E's, and the panic approach to managing. All are well-known barriers that prevent an organization from optimizing not only cycle time but also business performance. There are barriers that cause inordinate amounts of time to be lost without any benefit, and there are barriers raised by misconceptions, ignorance, and inappropriate implementation. The following barriers, listed in alphabetical order, include some that prevent organizations from dealing effectively with the three elements of cycle time (total time, timing, and cycle duration):

- The Baldridge Award
- Competition
- Delegating
- Incompetence
- Innovation
- Interfaces
- Japan Inc.
- Leadership
- Mistakes
- Transitions

The Baldridge Award

A 1987 Act of Congress created the Baldridge Award to encourage U.S. industry to improve the quality of its products and services. It has become the equivalent of the Academy Awards, the Pulitzer prize, and the Nobel Prize for business. The National Institute of Standards and Technology developed a 43-page set of guidelines that outline in detail the application process. It is ironic that U.S industry, which prides itself on private initiatives, espousal of free-market economic principles, and the latest in management expertise, needs the federal government to encourage it to achieve acceptable quality levels. On second thought, it is not only ironic but totally contradictory to the principles of industrial independence from government mandates. Does U.S industry really need the federal government to focus attention on the quality of its products and services? While we espouse individuality, we nurture conformity.

Main, in "Is the Baldridge Overblown?" [1], raises some major issues. The Baldridge Award does not provide any guarantee that an organization's products are superior to a competitor's. He states, "quality professionals were taken aback when Cadillac won the award in 1991." The award does not mean that Cadillac makes the best cars. It means only that the judges concluded that Cadillac has a quality system that meets the requirements of the Baldridge Award. Cadillac has developed the quality process. That may or may not affect the quality of the product. *The Wall Street Journal,* in a headline "All That's Lacking is Bert Parks Singing, 'Cadillac, Cadillac'" [2], illustrates the zealousness with which some organizations are pursuing this coveted public relations prize. However, the winners of the Baldridge have not necessarily resolved their business problems and gone, as Main notes, "to capitalist heaven."

Rohan, in "Do You Really Want a Baldridge?" [3], tells about life after the Baldridge: "If you win, it means speeches by the hundreds and guests by the thousands." Some examples:

- Globe Metallurgical Inc. (first small company, 210 employees) 25 to 50 calls per day
- Wallace Co. Inc. of Houston, 50 calls per day
- IBM Rochester, Minnesota, 270 calls within weeks
- Cadillac, 160 calls within weeks
- Federal Express, nearly 1000 calls
- Xerox, a four-person office to handle the requests

The Xerox story was told over 500 times in 1990 and heard by 55,000 to 65,000 people through speeches or visits. Xerox received 30 calls per day. Motorola has an office similar to that of Xerox: 735 requests in 1990, over 500 speeches, exposure to 10,000 people and 5400 organizations that have attended a monthly meeting or taken a course at Motorola University.

Garvin, a former member of the board of overseers of the Malcolm Baldridge National Quality Award, in "How the Baldridge Award Really Works" [4], states that the Baldridge Award, "has become the most important catalyst for transforming American business." He takes issue with those who claim it can be bought: Xerox a winner spent $800,000 and Corning, a runner-up, spent 14,000 hours in preparation. Garvin attempts to dispel the three main criticisms or myths about the award:

- Large expenditures are required
- It is flawed because it fails to take into account financial performance
- The award does not honor superior product or service quality

A practitioner's response to Gavin's rationalization from an operational perspective may be quite different. The effort expended by Xerox and Corning is substantial. According to *Business Week* in "The Ecstasy and the Agony," [5], overhead at Wallace Co. Inc., the small business winner in 1990, increased to $2 million per year, customers balked at higher prices, and by the time the company received the award, sales were slipping. The company lost $691,000 on revenues of $88 million, and the Maryland National Bank cut its revolving line of credit. Financial performance may not be a top priority, but it cannot be discounted. Garvin claims that the award is geared to future profitability. However, businesses are sustained by making some minimal return on investment. Regarding the third myth, arguments can be made that the scope is too narrow or too broad. Somehow the award must direct attention to the quality of the final product. Going through the process is insufficient.

Many organizations are now attempting to use the Baldridge guidelines to develop their quality processes. The award is treated as a fever or a virus: Take a Baldridge pill to reorient the organization to the necessity of improving quality. That appears to be overkill and the birth of another new management fad. There is no question about the importance of quality. The foundations for quality management are not limited to process: Process without substance, creativity, and innovation is wasted effort. Focusing on the process alone does not assure quality products or services. The issues that are examined in the Baldridge Award are valid. They can be guidelines, but as organizations attempt to institutionalize the approach, they only institute a new rigid process that may or may not provide a quality product. The continuous improvement approach to managing does not lend itself to such prescriptions as the Baldridge process. Is U.S. industry so depleted of creativity that it needs the federal government to induce it to produce products of acceptable quality by awarding prizes? Gold stars on report cards should be awards for kindergarten children.

Competition

A scan through any management journal shows the emphasis on managing in the 1990s and the use of "competitive," "competition," "competitor," or some combination with the appropriate noun or adjective. The literature stresses what others are doing. There is an overemphasis on—bordering on an obsession with—what the other company is doing. That applies to most papers presented at management conferences. It seems that every speaker begins with a statement about the inability of U.S industry to compete in the global marketplace. There is no doubt that knowing about the competition is important, but I suggest that organizations spend too much time being concerned about the competition and an insufficient amount of time looking inward to how the organization really functions and tapping its own creative talent. How valuable is this competitive information and how often does it affect business decisions? Should the intermittent and often erratic and desperate actions of even a formidable competitor change an organization's strategy? Why do intelligent managers conclude that their competitors are smarter than they are? There are no general answers to such questions.

Much of the information regarding the competition comes from an excessive amount of number crunching with the intention of simulating or modeling the competitor's strategy and operations. Simulation and modeling in business are not the same as simulations in science and engineering. In spite of the number-crunching capability, it is impossible to consider all the interactions that take place when a product is introduced. Number crunching does not generate new product ideas; it does not implement new manufacturing technologies; and it does not open new markets or identify the factors that may be needed to dominate a particular market niche. The success of any management activity can be described mathematically by the equation:

$$y = a_1 \times a_2 \times \cdots \times a_n$$

where y is the dependent variable and the a's are the independent variables. In other words, y depends on the activities defined by the a's. If the assumption is made that the a's are constant or independent, the management equation is ineffective. In the real world, there are thousands of a's and many of them interact. Some are supposedly inconsequential but have a major impact. Solving such an equation is not going to provide any significant value. Management has not yet reached the stage of a science whereby experimental results can be used for predicting future performance with any accuracy. Number-crunching data generate a great deal of mental noise that often seems to dominate the decision processes. That mental noise level regarding competition has surpassed even the decibel levels generated by the hard rock groups.

Without trying to explode the preoccupation with the many different facets of competition, I suggest that all the competitive effort be placed on the back burner and that organizations first learn to compete with themselves: continuous improvement throughout the complete organization. What was good yesterday must be improved today. Knowing what the competition is doing or projecting what it might do is important, but it sends the wrong message to the organization. The message is that we want to keep up with the competition.

Why not just promote the concept of continuous improvement because that is the thing to do? It provides the challenges to the participants and in the process allows the business to grow. Would the Ford Model T be on the streets today if competition never existed? Henry Ford would not have been satisfied with that level of performance. The telephone system of the 1940s was a virtual monopoly of AT&T. The company developed and applied new technologies not because there was competition, but because it provided growth. Someone had a vision of the possible. To sustain its research laboratories, AT&T needed continued growth to regenerate and provide better customer service and in turn more income.

Organizations that follow a catch-up strategy tend to emphasize the need for competitive information. It is difficult to find examples of successful organizations that follow the catch-up strategy. Playing catch-up with global competitors is not good enough: The competitors must be surpassed. That should not be confused with the strategy of being number two and waiting until some other organization takes a battering in the marketplace for technological or market reasons. Organizations that play the catch-up game seldom have time to get out in front of their competition. They spend all their time in a crisis situation attempting to stay alive. Waiting for the competitor to introduce a new product does not provide for business growth. The only realistic solution is to replace current products or render them obsolete.

The concept of replacing or rendering current products obsolete applies to all the functions in the organization: It can be recast as the concept of replacing all current policies and practices and rendering them obsolete when they are no longer applicable. U.S. productivity did not deteriorate overnight; the process has been going on for at least three decades. The difficulty lies in the fact that we continue to use the same tools but those tools are no longer applicable. Organizations that dominate a market and are at the head of the pack do not have to worry about the competitors. They just have to continue obsoleting their own products and understand the needs of the marketplace. IBM dominated the computer market. Control Data dominated a segment of that market. Eastman Kodak dominated the amateur film market. These were blue chip companies that grew and produced excellent results as long as there was no competition. They also forgot what it means to compete with themselves internally. They forgot about the need for continuous improvement. Those causes of trouble come from within and not from outside competition.

Delegating

The art and practice of delegating is one of the fundamentals of managing. The phrase "art and practice" is used intentionally because that is what it is. The fundamentals can be taught like any other subject, but knowledge comes from practice. Regardless of the length of time a person is in a managerial position, there is always something new to learn. The prerequisite is to recognize the uniqueness of people. That may sound like a cliché, but managers who practice managing understand the importance of taking uniqueness into consideration.

Managers who cannot delegate should be relieved of their responsibilities. Delegation does not imply a lack of participation and involvement by the manager, on the contrary, it provides the manager with the necessary time to get involved.

The *delegating equation* consists of three terms: delegating, follow-up, and feedback. Delegating without follow-up and feedback can have disastrous consequences. Dowd, in a *Fortune* article "What Managers Can Learn from Manager Reagan" [6], typifies much of the writing related to delegating. The Reagan history shows that although the president delegated, the follow-up and feedback never occurred. Perhaps too many executives took comfort in the *Fortune* article. Delegating is not a difficult task. The difficulty lies in knowing what and when to delegate, and then following up to make sure that the expected results are achieved. The delegation from a CEO does not go directly to the performer. It may go through several levels before it is actually accomplished. The follow-up is probably more important than the delegation. Why do executives hold reviews? To try to find out what is really taking place. If the act of delegation were the only part of that equation, executives would have no need to concern themselves about results. The assumption that all delegated work will be completed in a timely manner is erroneous and naive. Obviously, the degree of follow-up varies from person to person, experience level to experience level, and confidence level to confidence level, but that follow-up cannot be ignored by managers at any level.

Follow-up is not a matter of lacking in trust; it is just good business practice. The perfect communication link does not exist. Two groups are involved in the communication process: the transmitters and the receivers. The transmitters come from one side of the podium and the receivers from the other. It is important to recognize the differences. Executives are now paying the price for the hands-off type of management. Delegation involves balancing the hands-on approach so the system stays in control and at the same time provides individuals with learning opportunities. That involves risk taking on the part of the manager, and the risk taking involves the possibility that something may go wrong. That is part of a manager's responsibility: The manager takes the blame, and the employee takes the credit.

Incompetence

Incompetence is rooted in business culture and is found on all levels of the organization. It can infect an organization before managers even recognize its existence. It is an infection that slowly yields lower levels of performance regardless of the amount and type of medication. At the same time, everyone beginning with the CEO continues to think that the organization is in superb condition. This phenomenon begins with the rationalization of nonperformance. When organizations begin to believe the bottom-line figures and ignore the processes by which those figures were achieved, incompetence has already penetrated the organization. Management must ask itself why the competent become incompetent. Why has the organization lost the vitality that at one time was so evident?

Incompetence also results from acceptance of routine behaviors: It has worked so far, so leave it alone. That type of attitude helps to build what Argyris [7] described as skilled incompetence. It is the normal accepted behavior of the culture, and it is rewarded. Culture determines the skills that are used, those that are rewarded, and those that will be suppressed. Developing skilled incompetence could be a form of competence if that is what the organization rewards. Incompetence or obsolescence is generally associated with the scientific and engineering communities; obsolescence occurs because technologies are changing. The old and not so old are superseded by the new. That incompetence problem occurs in all business functions. None are exempt. The financial departments of many organizations are suffering from a lack of competence—not competence in putting numbers together, but in understanding the significance of the numbers. Too often, graduates in accounting become highly paid bookkeepers. Most are, at best, marginally literate in technology, marketing, and the realities of the business world. The story is told of the chief financial officer who recommended eliminating the customer service department as a means of reducing costs. Had he presented a plan of how the customers could be served after the function had been eliminated, he would have demonstrated a level of competence. His competence may have been in numbers, but he lacked business acumen.

Are the current unsatisfactory financial results (1991) due entirely to an economic downturn? Certainly the economy has an effect, but management must take a close look at the competence levels throughout the organization. As Americans, we like to boast of how we accomplish extraordinary things with ordinary people. However, within the ordinary group of people are the sparkplugs that keep the business engine running. Perhaps we have become too average. Perhaps we have reached the lowest common denominator. Competence must be continually upgraded and renewed.

Peters [8] proposes that organizations, as a matter of policy, subcontract everything and that they conceive of themselves as "nothing more" than a web of subcontractors. He quotes a senior technology company executive: We subcontract damn near everything, including most R&D. We must, to get beyond the confines of our own brain-dead company. That organization has lost not only its technical competence but also its managerial competence. The management also must be "brain dead" to have allowed such a condition to develop and then let it continue to exist.

Innovation

Innovation is not a predictable event. There are no specific schemes or rules that guarantee success. The academic researchers in the schools of management have not uncovered the secrets to effective and continuous innovation. Studies show that a specific type of environment is essential and that the infrastructure must be in place, but no theory that can guide the process has evolved. Even the research of Van de Ven, which began in 1982 and included a team of 15 professors and 19 doctoral students studying 14 different technology, product, and administrative innovations in the public and private sector [9,10], provides no significant input to the development of a theory of innovation. That does not imply that the work should be discounted, but the process of innovation is complex and does not occur as a result of a specific set of stimuli. It is a function of the individual.

Shani and Sexton, in "Myths and Misconceptions about the Dynamics of Innovation" [11], say that although innovation has been studied from a variety of perspectives (attempts to define, dissect, typologize, characterize, and so on), a significant amount of the knowledge about innovation is grounded in socially constructed reality. They suggest that organizations, managers, and innovators create myths that prevent understanding the process and that managers must move beyond the models of innovation and continually explore the dynamics within the organization that lead to innovation. The five myths are:

- *Definitional myth.*   If we can define innovation, we can systematically categorize the related activities and develop a theoretical base.

- *Success myth.*   The amount of innovation differentiates the successful from the unsuccessful enterprises.

- *Location myth.*   Innovation takes place within a particular type of physical and functional setting.

- *Organizational culture myth.*   Because innovation is a socially constructed process, it is likely to occur in a certain type of organizational culture.

- *Risk myth.*   The belief that the company that takes the biggest innovation risks will realize the highest payoffs.

Knox and Denison, in "R&D Centered Innovation: Extending the Supply Side Paradigm" [12], build on the work of von Hippel that identified two paradigms as alternatives to the initiation of the innovative process: the manufacturer-active and the customer-active paradigms. They confirm the findings of von Hippel. Limiting the initiation of the innovative process as described by von Hippel also limits the possibilities of being innovative. As a practitioner, I suggest that the sources of innovation are unlimited.

Personal experience has clearly shown that innovation does not begin with meetings or groups. It begins with a single person whose mind may have been triggered by some specific event or thought and who then pursued the idea relentlessly to some conclusion. The specific events, associations, pictorial compositions, and other stimuli that trigger the thought processes are unknown. Why did the inventor of 3M's Post-it notes come to the conclusion, at the time that he did, that he needed some means of tracking the pages in his hymnal? It was not the first time he sang in the choir. It was not the first time he faced the same problem. Logic would dictate that he should have thought about it at some time in the past. Some series of events that took place triggered the idea. It is questionable whether innovation can be programmed. Van de Ven talks about innovation teams and innovation managers, but those designations are inconsistent with reality. The manager and the team are a requirement, but usually after some one individual has formulated and explicated the concept. Someone had to originate the idea with a capital I.

Peters, in a two-part article "Get Innovative or Get Dead" [8,13], states that he has become obsessed with innovation. He raises a major issue that should concern all managers: the rigidity of approaches that perpetuate the idea that *one size fits all* and that innovation comes about as an orderly process or from the right company at the right time. Peters assigns the failure to innovate as one of the major causes for corporate deterioration in performance. He believes that innovation must be put at the top of the business agenda beginning with what he describes as "violent market injection strategies."

Much of the controversy surrounding innovation arises from the fact that the word "innovation" means different things to different people. Is it invention plus commercialization? What is the difference between an innovation and a product extension or enhancement? 3M's Post-it notes were originally introduced as little yellow notepads. That was the innovation. The extensions to different sizes and colors, imprinted and die-cut versions, personalized and so forth, are routine product or technology enhancements. They are only logical extensions of the original technology; they are expected to be introduced through normal business operations. Not every configuration of the basic design is an innovation. Logical and continuous technology and product extensions are just good business practice; they tap new and different segments of the marketplace for added value with only incremental investment of resources. That is leveraging of business resources. An innovation is more than an invention; it includes the commercialization.

There is no argument about the necessity for U.S. organizations to become more innovative not only in matters related to new products and processes but to the total business. Innovation must be pursued throughout the organization. Managers must translate concept to commercialization to concept to implementation. Not every innovation must generate a new business. Raising expectations may be the best innovation that managers could pursue, beginning with themselves.

Interfaces

Managing the many business interfaces presents one of the most formidable challenges to managers. The interfaces are both internal and external and involve two distinct and major considerations:

1. Transfer of information (technology)

2. Developing effective working relationships

The more generalized term "transfer of information" is used in preference to "transfer of technology" because managing cycle time involves more than technological information. There is ancillary information that cannot be disregarded. It originates in all the other administrative functions. Technology transfer is a subset of information transfer. It must be supported by the information coming from the administrative functions in order to provide a benefit to the organization. Customers do not buy technology. Businesses do not (should not) invest in technology for its own sake but for how it can be used to exploit the available resources.

Brown [14] describes the issues related to managing interfaces and their importance, as businesses come under more pressure to innovate in an ever-increasing competitive global economy:

> Organizational interfaces and the conflict problems within them will be an important part of the future. Technical specialization, power inequities, cultural diversity, and organizational independence are continuing—often expanding—realities of life in complex societies. Managing differences creatively at the points where social units come together will continue to pose challenges. Inventors of innovative and effective responses to these challenges will be critical assets to their organization—and to the larger societies in which they live.

Gaynor, in "Achieving the Competitive Advantage through Integrated Technology Management" [15], includes a discussion of managing interfaces in relation to the management of technology. The topics include the role of inter- and intrafunctional interfaces, the people characteristics, the influences of single-issue management, the ability of managers to integrate the activities of people by focusing effort, the need to create dissonance, criteria for measuring impact of interface mismanagement, organizational structure, and education of the participants. Managing interfaces of all types and at all levels is a major business issue. Those interfaces include:

- Research and marketing
- Research and design/development
- Research and manufacturing
- Design/development and marketing
- Manufacturing and marketing
- All administrative functions

M. F. Wolfe, the editor of *Research · Technology Management* [16], calls attention to the dual interface—the transfer of technology from R&D to manufacturing—in an interview with Keith McHenry. At the time (1985), McHenry was vice president for R&D at Amoco Oil Co. He states that at Amoco manufacturing has an effective interface with R&D because:

- Researchers work on a project basis and every project has a sponsor either in manufacturing or marketing.
- Research required for technical support is financed from the manufacturing budget.
- Research people are encouraged to participate in plant start-up activities and to consult with manufacturing.
- A solid communication network exists between the technical specialists in engineering, R&D, and the line organization in the plants.

The interfaces between design and manufacturing are considered by Dierdonck in "The Manufacturing/Design Interface" [17]. There is little argument that the interface must be mediated because of its impact on time-to-market, but Dierdonck raises issues about the limits of integration when major or radical innovation is involved. Innovations with close linkage may of necessity be treated differently. This is just another case, though, as suggested in Chap. 8, of differentiation and integration lying on a continuum and it is the manager's responsibility to make the appropriate decisions.

Most of the research has not shed a great deal of light on the linkage between organizational practice and productive research-marketing interfaces. Research complains about the quality of the marketing input, and marketing complains about the nonresponsiveness of the research community to market needs. There is generally little credibility between these two functions and little understanding of the contributions expected from each. Gupta and Wilemon, in "The Credibility-Cooperation Connection at the R&D-Marketing Interface" [18], combine the issues of information and source credibility, organizational practices, and cooperation to analyze R&D's perception of marketing input in companies that had different levels of integration.

In the Gupta-Wilemon study, marketing information was perceived to be credible if it is realistic and valid, well analyzed and presented, objective, consistent and complete, useful, and appealing. The marketing manager was perceived as credible if he or she was cooperative, open and trustworthy, competent and helpful, friendly and social, fair and easy to work with, knowledgeable of R&D, a rational decision maker, and respected. Nothing was mentioned about focusing on results, being proactive, looking to the future, and being intuitive, a risk taker, and so on—all qualities that are needed for success. Credibility is a determining factor in business success. It applies to all interfaces whether related to individuals or departments. Once a person loses credibility with colleagues, it is a long and tiresome grind to regain it. If organizational practice, beginning with the CEO, does not support credibility and cooperation, interface problems are exacerbated. Credibility and cooperation are not choices; they are the rights of a supportive partnership and a management that seeks to balance the needs of the individual with the needs of the organization.

Lucas and Bush, in "The Marketing-R&D Interface: Do Personality Factors Have an Impact?" [19], raised the issue of personality differences between research and marketing managers. They focused primarily on 16 personality factors to determine the significant differences between marketers and researchers. As one would assume, there are significant differences. They found that "marketing decision makers were more dominant and assertive as well as more happy-go-lucky and enthusiastic than their counterparts in R&D." The list of differences continued with marketers being more venturesome and spontaneous and more group-dependent, whereas individuals in R&D were more self-sufficient. The factors relative to intelligence, ego, conscientiousness, and so on, were present in both groups. The research does show that marketing and R&D managers are different on personality dimensions that are important to their respective positions. They also raise the issue of who should dominate product development, research and development, or marketing. As a practitioner and observer for almost 40 years, I found that if either dominates through excessive influence, the organization faces potential problems unless it has that ideal blend of marketers who understand the technology and its limits or researchers who understand the needs and idiosyncrasies of the marketplace.

Rosenthal, in "Bridging the Cultures of Engineers: Challenges in Organizing for Manufacturable Product Design" [20], relates the conventional view of manufacturing engineering as described by Koenig [21]. Koenig divides the field into four functions: advanced manufacturing engineering, process control, methods planning and work methods, and maintenance. He states that:

> A separate engineer could have each or several of these responsibilities, or several engineers could share one area of responsibility. It matters little what the specific organizational structure looks like as long as all of the responsibilities are attended to.

Rosenthal takes issue with Koenig relative to the role of organization. He suggests that organizational structure determines the culture of manufacturing. If no specialists are formally assigned to investigate the manufacturability of a new design, then manufacturing will only focus attention on improving the existing processes. Rosenthal considers another scenario that includes an independent group of "producibility engineers." He describes producibility engineers as:

> A producibility engineer...uses as inputs designs from design engineering and capabilities observed in the factory, then converts those designs and capabilities into workable designs so they can be made in the factory. The producibility engineer, then, is a compiler of information, an optimizer of factory input and design input into a producible scheme.

Rosenthal goes on to suggest that the primary function of the producibility engineer is to initiate the design reviews at the time that a design engineer is ready to present drawings of a part or an assembly.

Resolving the design-to-manufacturing interface by appointing a producibility engineer loses sight of the nature of the problem. It adds another individual or group to the process flow and further stratifies the organization: a group to do something over again when it has already been done. Eliminating the problems between design and manufacturing does not need more people. It requires that design engineers be adequately educated about the means for producing their designs. The charge is often made that designers choose elegance of design over practical manufacturing considerations. Design departments usually operate under the leadership of a design manager. If the manager cannot control the creeping elegance, then he or she must be either reeducated or replaced.

Team building

Eliminating interfaces involves developing some form of team approach. Projects are organizations of people into teams. As was stated, organizations meet their financial targets through the use of some form of team. Team building at the professional levels has received considerable attention as a means of decreasing the negative effects created by the specialized and segmented functions. Verespej, in "When You Put the Team in Charge" [22] notes that management believes that self-directed work teams are going to accomplish miracles. The first point that must be recognized is that teams are not another management panacea. They are applicable in some situations but not in others. They can enhance or hinder an operation depending on the team members and the purpose for which the team was established.

Self-directed teams emphasize the sharing of personnel functions, which include selecting team members, evaluating the individual performance of team participants, making compensation decisions, determining the educational needs, setting the pace of the work effort, and so on. In addition to those personnel issues, the teams pursue the normal activities associated with establishing the project goals, implementing all the requirements of the project, and the many other activities of any project organization. Verespej provides some insight to the use of self-directed teams and some of the factors that influence their performance. A survey conducted jointly by the Association for Quality and Participation and *Industry Week* revealed that organizations underestimate the financial investment in worker education required to help self-directed teams succeed. More than 70 percent of the effort is directed at manufacturing, and the average team size of the respondents was six to ten people. The survey also reports that self-directed teams benefit from team involvement and performance, positive morale, a sense of ownership and commitment, improved quality, and productivity.

Self-directed teams are not without their problems. Education is one vital ingredient for success, and the linkage between business strategy and a vision of where the organization is going must be clearly established. The survey also indicated that self-directed teams have their share of personality clashes.

Teams get to be teams by learning from one another and recognizing the need for the expertise and the contributions of others to meet certain objectives. The symphony orchestra represents the ultimate in teams: professionals under the guidance of a director playing from the same score but different sheets of music. Developing a team that adds value to the enterprise requires more than a superficial approach to appointing the participants.

Cox, in an article "The Homework behind Teamwork" [23], provides a comprehensive description of the value-adding team:

> The value-adding team is a collective state of mind where ideas are food and puzzlements are challenges. It is where conflict is positive because it is out in the open. Responsiveness is paramount. Whether established department or ad hoc force, *team is a thinking organism* where problems are named, assumptions challenged, alternatives generated, consequences assessed, priorities set, admissions made, competitors evaluated, missions validated, goals tested, hopes ventured, fears anticipated, successes expected, vulnerabilities expressed, contributions praised, absurdities tolerated, withdrawals noticed, victories celebrated, and defeats overcome. Finally it is where decisions are backed when the boss says yes or no to a particular option.

That description does not specify any type of structure. As a matter of fact, if a team functions according to Cox's description, organizational structure is unimportant. As was stated previously, if the right people are available, structure is of little if any importance; without the right people, no organizational structure will produce the expected results.

The Wall Street Journal reports that some big companies are attempting to speed the decision-making process by eliminating the interfaces through appointment of "innovative senior management teams" [24]. According to Nadler of Delta Consulting Group, few operate autonomously or have any significant impact on the entrenched corporate hierarchy. He states that Kodak's chemical division replaced three senior vice presidents with self-directed management teams without a designated leader: one for R&D, one for manufacturing, and one for administration. The divisions president suggests that "decisions get made faster because there are fewer layers."

Breaking down the barriers that inhibit effective and efficient transfer of information will take time, cost money, and require business discipline. It will demand building the operational infrastructure; it will require that managers function as managers rather than as hired hands or storekeepers; and it will demand that managers recognize their role in making a creative contribution to the business. The action begins with the CEO. A commitment—visible, real, and actively participative—is a prerequisite for eliminating the interface problems. The objective must be the elimination of all interface problems between individuals and groups. Meeting such an objective requires a change in the state of mind by every person in the organization. In some cases, some strong measures may be needed to change the organization.

Developing an effective team in an organization in which 15 or 20 individuals develop into a well-synchronized group of professionals may be a start to resolving interface problems, but unless the organization functions as a team, only a minimal benefit will be obtained. The metaphor taken from the sports world must extend beyond the playing field. The sports team includes more participants than those on the playing field; many other people are involved in allowing a relative few to demonstrate their talents. There is a system behind the metaphor, and that system must be satisfied if an event is to be successful. If the concept of the *team* has any significance whatsoever in business, the business must function as a team. That requires following the description provided by Cox.

Japan Inc.

Japan bashing continues not only on Capitol Hill but in the executive suites in U.S industry. In reality, it consumes time that might be better spent looking inward and facing up to the inefficiencies within the organization. Rationality has not dominated the political scene. Industry leaders who consider themselves rational have become emotional about Japan Inc. At first, many industrial leaders thought that Japanese methods could be transferred to their businesses. In the process, they first forgot that the methods of one culture cannot be transferred to another culture and then they only wanted to transfer some methods. More important, they did not understand Japanese culture or the Japanese approach to managing. They jumped on the quality-circle bandwagon and other fads that went the way of all fads. The media exploited the situation, and everything that was Japanese was "good" and everything that was made in the United States was "bad." The truth was somewhere in between the two extremes. Smith, in a *Fortune* article, "Fear and Loathing of Japan" [25], presents the results of a survey of what Americans tell the poll takers about Japan. The negative attitudes have gone beyond the blue-collar workers. As the *Fortune* article notes: An intellectual patina comes from the so-called revisionists, a school of journalists, academics, and others who argue that Japan is neither righteous nor wicked, just different and will never have a free market like the United States, so the only answer is some form of managed trade.

Industry should recognize that Japan is here and that the bashing and the rationalization of poor performance in U.S. industry, as compared to the Japanese, will not improve the effectiveness and efficiency of U.S. industry. Japan Inc. threatens U.S. industry because U.S. industry is looking for solutions to the wrong problems. Trade relations are a real concern. Emphasizing the external factors instead of facing up to the real issues related to productivity will not make U.S. industry more productive.

Made in America [26], presents the results of an extensive study of U.S. industry by the Massachusetts Institute of Technology (MIT). MIT undertook the task: "to identify what happened to U.S. industrial performance and what we [MIT] and others might do to help improve the situation." The results are presented under the following major chapter headings:

- Outdated strategies
- Short time horizons
- Technological weakness in development and production
- Neglect of human resources
- Failure of cooperation
- Government and industry at cross-purposes
- Emerging patterns of best industrial practice
- Imperatives for a more productive America
- Strategies for industry, labor, and government
- How universities should change
- Conclusions

The studies specifically related to the automobile, chemical, commercial aircraft, consumer electronics, machine tool, semiconductor, computer, copier, and steel industries.

The conclusion of the study is summarized in less than two pages and directs industry's attention to the following four conclusions, which are quoted directly in some cases and paraphrased in others:

- The United States does have a serious productivity problem. The problem of "productive performance" began in the past four decades and has manifested itself in sluggish productivity growth, shortcomings in quality, and lack of innovation.

- The causes go beyond the macroeconomic explanations. The weaknesses are deeply rooted and affect the way people and organizations interact with each other and technology; they affect the way businesses deal with long-term technological and market risks; they affect the way educational institutions go about developing the nation's most precious asset, human resources; and they introduce rigidities into the nation's production system at a time of extraordinarily rapid change in the international economic environment.

- Although some U.S. firms have adapted to the new environment, the diffusion of what have been identified as best industrial practices is inhibited by ingrained beliefs, attitudes, and practices. Addressing yesterday's weaknesses offers no immunity from problems caused by tomorrow's changes.

■ Although there is no cause for despair, there is also no cause for celebration. The United States has three great strengths: creativity, entrepreneurship, and the energy of individual Americans. If industry, government, and the educational system can unite and build on those strengths, we should be able to enter the twenty-first century with the same dynamism and leadership that characterized industrial performance throughout much of this century.

Our threat is from within our own organizations, not Japan. It is easy to blame the Japanese for the problems created for U.S. automakers. But who created the problem? Restriction on imports will not make a better-quality car.

Leadership

Leadership [27,28] by itself may be of academic interest, but in the world of business it must be linked with a vision of the direction in which the organization is moving. It is a managerial imperative. The vision is not an idle dream or fantasy; it is based on knowledge, some speculation, and faith in people who have demonstrated their ability to act. Leadership cannot be equated with dollars in the business treasury. It requires providing for today and tomorrow. The whole concept of leadership means creating change as contrasted to maintaining the status quo. It implies thinking of the future, influencing, persuading, changing minds, doing what those above and below may consider to be unacceptable, sticking your neck out, taking calculated risks, risking yourself as a person in championing a controversial point of view or approach, and having the confidence and ability to speak out and support unpopular but necessary issues. It goes beyond playing king of the hill.

Badawy, in *Developing Managerial Skills in Engineers and Scientists* [29], emphasizes the need for managerial knowledge, skills, and attitudes in achieving managerial competence. Knowledge is expertise. In the preface to his book, Badawy says,

> Like engineering and medicine, effective management requires both knowledge and practice. Knowledge without practice breeds a 'blue sky theorist.' Practice without knowledge breeds a trial-and-error layman. Knowledge and practice breed a well grounded, competent practitioner.

Managing involves leadership and leadership is not limited to CEOs. On the contrary, the CEO's contribution to leadership may only involve providing the environment in which others demonstrate their leadership—leadership in that vast group of professionals who conceive the products, develop them, manufacture them, and service the customers.

Managing involves three distinct activities: administration, providing direction as contrasted to directing, and leadership [30]. All are required in the correct amounts for business success. The manager who focuses only on administration is a bureaucrat. The manager who provides only direction, although more productive than the bureaucrat, cannot achieve all the potential for success. The manager who thinks that all of his or her time can be spent leading also will be ineffective. All three activities are part of managing and must be applied in varying degrees depending upon the specific situation.

Zaleznik in "The Leadership Gap" [31] raises the critical issues of leadership. He contrasts the role of the leader with that of the manager and states that "business executives erroneously believe that management and leadership are synonymous; that to manage according to the principles of its mystique is to lead." Leadership involves going beyond the current body of knowledge or accepted facts, processes, and rituals. Leadership requires stepping out of line and doing the unpredictable. Following a process and hoping that it will produce the expected results is not leadership. Administration is the more appropriate word for that and, as mentioned, any manager who spends all of her or his time in administration cannot succeed as the dynamics of the business force a change in direction.

Two quotations from Zaleznik sum up the answers to leadership. They not only support my particular biases, they have been observed as the differences between leaders and managers. In the first quotation, Zaleznik argues that:

> Leadership is an argument for substance over process that calls for restoring the individual to his or her proper place as the source of vision and drive that can make an organization unique.

In the second and equally important quotation, he goes on to say that:

> Professional managers with their bag of tools can indeed coordinate and control, but have lost sight of the substance of work in business. Self-esteem follows, not from submerging oneself in the team and following process, but from facing problems, assuming responsibility, and doing good work.

Making Mistakes

Peters has been preaching the gospel of "making mistakes" [32]. Making mistakes may be important, but it must be put into an operational perspective. First, consider the cost of adopting fads as a management style, that is, "managing with fads." Managing with fads is what I have often referred to as "pursuing the momentary triumph" until the next one comes along. Gimmicks and the quick fixes do not resolve business issues. The tragedy of the concepts that become fads is that all have some merit and substance. Most of the fads, as shown in Fig. 9.1, fall into two categories, those associated with people and those associated with things.

The two lists in Fig. 9.1 represent the more common fads of recent origin. Scientific management has not been included because most management is not scientific, although the principles may apply. Can anyone argue that any of the fads in both lists cannot provide some advantage? Each represents a major factor in improving business performance, but each is also short-lived. The difficulty arises when each of the fads is considered independently as a panacea for solving problems. The difficulty is exacerbated when the gurus present their quick-fix courses without understanding the underlying principles and the requirements for implementation. How much did executives really understand about the process of strategic planning before they introduced it? Not much. Had they fully understood the limitations and the associated costs, they might have given more thought to strategic thinking and developing a strategy rather than a plan. Managers must focus on the concepts that turn into fads, but the limitations and interrelations with the business system must be understood.

People	Things
Theory X, Y, and Z	Artificial Intelligence
Management by objectives	Computer-aided design, inspection, manufac-
Employee involvement	turing, and integrated manufacturing, and
Job enrichment	so forth
Empowerment	Continuous-flow manufacturing
Team building	Total-quality management
Innovation and entrepreneurship	Just in time
Corporate culture	Organizational development
Value analysis	Zero-based budgeting
Quality of work life	Learning organization
Excellence	Zero defects
Participative management	World-class manufacturing
Sociotechnical systems	Total preventive maintenance
Literacy of all types	Work simplification
New paradigms	Group technology

Figure 9.1 Management fads or single issues classified by people and things.

This leads to the topic of discussing "mistakes." The gospel of encouraging and making mistakes has some limitations. First, it requires understanding just what is meant by making mistakes. As has been noted elsewhere in this book, mistakes in judgment are quite different from mistakes that are due to negligence or inattention to detail. Any experimental laboratory depends on making mistakes and dealing with failures to gain knowledge. The failures must rest on a firm foundation; they should not occur because of haphazard work methods or other controllable factors. Peters glorifies mistakes and failures. He says that:

Complexity + need for speed=make more mistakes (or else)

He continues by saying that: "There's little that is more important to tomorrow's managers than failure. We need more of it. We need faster failure."

It is questionable whether Peters espoused that philosophy when he was a partner in McKinsey & Co. There is no argument that mistakes will be made and that failure is part of making progress—part of the learning process. There is no evidence that there is any positive correlation between innovation and failure. As a career employee at 3M, I found that mistakes and failure are an accepted fact of business life, but to the best of my knowledge 3M management does not celebrate mistakes and failures and accept Peter's suggestion of: "Have an annual 'Hall of Shame' banquet where you give awards for the fastest/most useful—and the dumbest and most embarrassing foul-ups."

The questions related to mistakes, goofs, foul-ups, and failures are important business issues. Their importance relates directly to their size. Blowing $100,000 on an experiment, even if no new learning takes place, may or may not be important. It depends on what portion of the available financial resources are blown. To a billion dollar organization it means little; to a start-up it could be disastrous. To treat making mistakes and glorifying failures as candidates for a new fad diminishes their importance to furthering innovation. Innovation of any type requires risking resources. Mistakes will be made and failures will occur. Hopefully, some learning will take place in the process.

Recognizing rather than covering up failures and mistakes brings an organization down to reality. Not everything is going to succeed, and attempting to rationalize failures passes the wrong message to the organization. Why are so many organizations reporting lower earnings in a downward economy? The economy did not reverse itself overnight. The signs were evident, but most organizations did nothing to prepare for what was coming. Blaming the economy does not improve earnings; it does not improve effectiveness and efficiency. Why not confront the organization with the truth: mismanagement, lack of acceptable performance, absence of discipline, and various levels of apathy on the part of everyone beginning with the CEO and executive level management. It is their decisions and policies and practices that brought the organization to where it is. They appointed managers who became the storekeepers.

Transitions

Effectiveness of any group of individuals depends upon the ability to manage transitions. That applies to the technology and marketing operations as well as the other functional groups. Introducing new technologies involves transitions, and so does entering new markets. Introducing system cycle time management involves transitions. However, transitions involve more than knowledge, restructuring, and various degrees of uncertainty; the attitude with which they are approached is more important. On paper, restructuring is a simple process; making it work depends upon the attitude of the participants. The announcement of staff reductions of 20,000 people at IBM involves a transition that is evidently locked in the minds of the CEO and senior management. It is a top-down dictum based on financial performance without consideration of the managers who will be involved in making the transition. Although the need for such restructuring must begin with senior management, the sources of information for determining the changes that must take place must come from those responsible for implementing the final decision. It may be more prudent to withhold such announcements until a plan that delineates the specific changes that must take place has been developed.

John Akers, IBM's CEO, has been trying to change the direction of the company since 1988. According to *The Economist* [33], the proposals that were made in 1988 to decentralize the company, slim it down, and make it more entrepreneurial never materialized. The question that remains is whether the latest restructuring will make a significant difference. IBM must go through some logical transition plan with clearly delineated objectives if the 1988 failures are not to be repeated.

Transitions may be major or minor. Making transitions effectively involves understanding the conditions and the environment that will allow the transition to take place. As in other situations, making the transition involves doing the up-front work before making the decision. Figure 9.2 illustrates the process of making effective transitions. It shows the current state and two desired states, one above and one below the current state. The two desired states are used to demonstrate

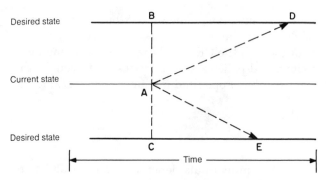

Figure 9.2 Simplified representation of making a transition from the current state to some new desired state.

that the desired state may not necessarily be better, but it will be different and will require some behavioral changes.

The major difficulty associated with a transition stems from the fact that the decision makers attempt to make the transition from one state to another in zero time, as shown in Fig. 9.2 from A to B or from A to C. The basic elements of the plan and the potential difficulties in making the transition are not delineated. That does not mean that all can be perceived and uncovered in the up-front work, but a large percentage can be. Making a transition requires a design. In the world of technology, attempting to perform any action in zero time tends to throw the system into oscillation. That type of action causes the system to either shut itself down or destroy itself. IBM's 1988 attempt to restructure, in essence, shut itself down. A more appropriate way is to determine what can be accomplished in a specific time with a workable plan and reach the desired state at D or E. Making the decision is simple; implementing it can be accomplished only with the support of the participants. Elementary. Yet two major corporations, IBM and General Motors, announce major restructurings dictated from the top. As an engineer, executive, and consultant, I have had experience in participating in turnaround situations. One thing is certain: The need to restructure may come from the CEO, but the focus of the new business unit or function must first be determined. The people issue cannot be resolved until the new directions are clearly identified. IBM's proposed staff reduction appears to be a financial decision based upon cutting people from the payroll but not knowing from where or from which programs.

Managing transitions, whether they involve the company, a function such as R&D, a technology, or a change in culture, can either enhance or inhibit performance. Making a transition to system cycle time management involves transitions in thinking, in attitude, in perspective, in approach, in scope, and in expectations.

Summary

Optimizing cycle time requires removing the barriers to effective and efficient management. Attempting to change the current bureaucracies through some mild and long-term evolutionary process will achieve nothing. To change requires specific actions. A revolution, although it may be necessary, requires the appropriate infrastructure and resources. The *new model* requires a renaissance in *thinking* about the management process. Emphasizing any single issue will not change the manner in which a business unit attempts to meet its objectives. The *new model* must consider all the elements of the business system. Some proponents of developing new business paradigms call for revolutionary changes, but they disregard the development of the environment for accepting the changes. The revolution must be created through a continuous series of minirevolutions. Creating a change and then institutionalizing it only develops new manifestations of rigidity and inflexibility that create impediments and raise future obstacles. Eliminating one set of barriers and substituting another serves no long-term purpose. The new model must focus on a philosophy of continuous improvement, because what is acceptable today will most likely not be acceptable tomorrow.

There is no doubt that resolving the negative aspects created by all barriers is complex and will challenge the best of managers. That complexity came about as a result of overspecialization, oversegmentation of business units, and lack of managerial competence. Much of the problem can be resolved by providing the necessary leadership and viewing the business as a system rather than as a collection of independent entities.

References

1. J. Main, "Is the Baldridge Overblown?," *Fortune,* July 1, 1991, pp. 62–66.
2. Yoder, Fuchsberg, and Stertz, "All That's Lacking is Bert Parks Singing, 'Cadillac, Cadillac'," *The Wall Street Journal,* Dec. 13, 1990, p. 1.
3. T. M. Rohan, "Do You Really Want a Baldridge?," *Industry Week,* April 1, 1991, pp. 11–12.
4. D. A. Garvin, "How the Baldridge Award Really Works," *Harvard Bus. Rev.,* November-December 1991, pp. 80–93.
5. M. Ivey and J. Carey, "The Ecstasy and the Agony," *Business Week,* Oct. 21, 1991, p. 40.
6. A. R. Dowd, "What Managers Can Learn from Manager Reagan," *Fortune,* Sept. 15, 1986, pp. 33–41.
7. C. Argyris, "Skilled Incompetence," *Harvard Bus. Rev.,* September-October 1986, pp. 74–79.
8. T. Peters, "Get Innovative or Get Dead, Part 1," *California Manage. Rev.,* vol. 33, no. 1, Fall 1990, pp. 9–26.
9. T. Brown, "Why Companies Don't Learn," *Industry Week,* Aug. 19, 1991, pp. 36–43.
10. A. H. Van de Ven, H. L. Angle, and M. S. Poole, *Research on the Management of Innovation,* Harper & Row, Publishers, New York, 1989.

11. A. B. Shani and C. F. Sexton, "Myths and Misconceptions about the Dynamics of Innovation," *National Productivity Rev.,* Winter 1990/91, pp. 75–84.
12. S. D. Knox and T. J. Denison, "R&D Centered Innovation: Extending the Supply Side Paradigm," *R&D Management,* vol. 20, no. 1, 1990, pp. 25–33.
13. T. Peters, "Get Innovative or Get Dead, Part 2," *California Manage. Rev.,* vol. 33, no. 2, Winter 1991.
14. D. L. Brown, *Managing Conflict at Organizational Interfaces,* Addison-Wesley, Reading, MA, 1983.
15. G. H. Gaynor, *Achieving the Competitive Edge through Integrated Technology Management,* McGraw-Hill, New York, 1991.
16. M. F. Wolfe, "Bridging the R&D Interface with Manufacturing," *Research Technology Manage.,* January-February, 1985, pp. 9–11.
17. R. Van Dierdonck, "The Manufacturing/Design Interface," *R&D Management,* vol. 20, no. 3, 1990, pp. 203–209.
18. A. K. Gupta and D. Wilemon, "The Credibility-Cooperation Connection at the R&D-Marketing Interface," *J. Prod. Innov. Manage.,* vol. 5, no. 20, March 1988, pp. 20–31.
19. G. H. Lucas and A. J. Bush, "The Marketing-R&D Interface: Do Personality Factors Have an Impact?," *J. Prod. Innov. Manage.,* vol. 5, no. 4, December 1988, pp.257–268.
20. S. R. Rosenthal, "Bridging the Cultures of Engineers: Challenges in Organizing for Manufacturable Product Design," in *Managing the Design-Manufacturing Process* by J. E. Ettlie and H. W. Stoll, McGraw-Hill, New York, 1991, pp. 21–52.
21. D. T. Koenig, *Manufacturing Engineering,* Hemisphere Publishing Corp., New York, 1987.
22. M. A. Verespej, "When You Put the Team in Charge," *Industry Week,* Dec. 3, 1990, pp.30–32.
23. A. Cox, "The Homework behind Teamwork," *Industry Week,* Jan. 7, 1991, pp. 21–23.
24. J. S. Lublin, "Managing," *The Wall Street Journal,* Dec. 20, 1991, p. B1.
25. L. Smith, "Fear and Loathing of Japan," *Fortune,* Feb. 26, 1990, pp.50–60.
26. M. L. Dertouzos, R. K. Lester, R. M. Solow, and The MIT Commission on Industrial Productivity, *Made in America,* The MIT Press, Cambridge, MA, 1989.
27. G. H. Gaynor, *Achieving the Competitive Edge through Integrated Technology Management,* McGraw-Hill, New York, 1991, pp. 58–60.
28. M. K. Badawy and G. H. Gaynor, "Mismanagement of Technology: An Issue Facing American Industry," in press.
29. M. K. Badawy, *Developing Managerial Skills in Engineers and Scientists,* Van Nostrand Reinhold Company, New York, 1982.
30. G. H. Gaynor, *Achieving the Competitive Edge through Integrated Technology Management,* McGraw-Hill, New York, 1991, pp. 211–220.
31. A. Zaleznik, "The Leadership Gap," *Academy of Management Executive,* vol. 4, no. 1, 1990, pp.7–22.
32. T. Peters, *Thriving on Chaos,* Harper & Row, Publishers, New York, 1987, pp. 315–320.
33. "Scenting Extinction," *The Economist,* Dec. 14–20, 1991, pp. 69–70.

10

Implementing a Responsive Management System

System Cycle Time Management (SCTM) must be developed within the context of the total business system. "Responsive management system" implies that management at all levels, through the other participants, will respond by managing total time, timing, and cycle duration as a unit. The level of accomplishment depends on using the seven elements of the infrastructure effectively and also using the ten business resources to pursue business-related activities. All the elements of the infrastructure must be in place, understood, and reflected in the everyday practices of the organization. Use of the resources must add value and leverage. This chapter considers additional requirements of a responsive management system:

- Mastering the human resource issues
- Managing the dynamics of the business
- Resolving the information paradox
- Dealing with discontinuities
- Measuring performance
- Expecting innovation from all functions
- Gaining advantage from organizational learning
- Managing the process while focusing on substance

Mastering the Human Resource Issues

Figure 10.1 shows the linkages among product genesis, distribution, and the administrative entities of the tripartite organization. The activities of product genesis and distribution largely determine the success of the business, and that applies to both product and service businesses. They are responsible for generating the largest percentage of the sales value

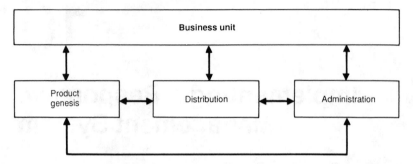

Figure 10.1   Major inputs and feedback loops in a responsive management system.

of production within an organization—in excess of 70 to 75 percent. They are also responsible for most of the cost associated with the sales value of production—generally in excess of 50 to 65 percent.

The relations and interdependence of people, leveraging of resources, interface issues, and selected value-adding activities in pursuit of some predetermined outcome have been established. *People* as a resource is singled out from the ten resources because people determine the level of optimization and the optimal use of the other nine resources. Consider a new-to-the-market product from the time the go-ahead to proceed is given. If the assumption is made that the resources are available to pursue the project, the success will depend totally upon the people component. The argument could be made that the economy or factors beyond the control of the people influences the outcome. Eventually, the final responsibility rests with people. People generate ideas and concepts, make the decisions, use the resources, determine the level of uncertainty and risk as they perform activities, people implement ideas and concepts, and misread the ups and downs of the economy. Whether people control all the unexpected interferences and uncertainties is not important. The depth and the accuracy to which they have analyzed the potential risks and synthesized the available information into a cohesive plan determines the level of success. Ultimately, level of performance depends upon people.

People are also responsible for creating and resolving the interface issues and leveraging the business unit resources. The degree to which value is added, through the multiplicity of activities, depends entirely on how the resources will be leveraged and the degree to which the walls between the various fiefdoms have been broken down.

Changes have been taking place in the industrial world, yet human resource management has remained static. Organizations have reduced cost; advances have been made in product quality; and product development times are being reduced. Those changes, as suggested by

Bounds and Pace in *Human Resource Management for Competitive Capability* [1], do not allow organizations to continue with the cultural comforts of the past. They emphasize the development of human resources at all levels based on the argument that well-developed human resources are the means to continuous improvement. Bounds and Pace accuse the human resource professionals of exhibiting technical arrogance and "functional specialization and functional isolation." Human resource managers, like many staff managers, believe that all problems can be resolved through departmental expansion.

A responsive management system must focus attention on all the elements that affect people performance. A responsive human resource policy requires eliminating much of the embedded entitlement attitude engendered in organizations during the past several decades. It calls for a new reality that focuses attention on raising the expectation and performance level of every participant beginning with the CEO. Meshoulam and Baird, in "Proactive Human Resource Management" [2], present a model for anticipating and responding to the human resource management needs of a business. They suggest that human resource management is "living in the past" and that practices and procedures that have been historically developed from trial and error have lagged behind the organizational needs.

There is no doubt that human resource departments impact an organization's culture—unfortunately and too often, negatively in regard to performance. The article includes an in-depth discussion of the stages of development of a human resource department, but it fails to consider the impact on people. The authors quote a human resource professional, "We are becoming better and better at running systems and programs that are less and less relevant."

Baird and Meshoulam, in "Managing Two Fits of Strategic Human Resource Management" [3], consider the need for external as well as internal fit. Structure, systems, and managerial practices and procedures must fit the organization. The needs of an organization change over time, even in the area of managing human resources effectively. The internal fit involves the different components associated with human resources. The authors identify five stages of human resource management: initiation, functional growth, controlled growth, functional integration, and strategic integration. Human resource departments either enhance or inhibit the performance of the technology-oriented functions and participants, depending on the breadth of their outlook. They exist to make a contribution to the business. They must become active participants.

Compensation at all levels in the hierarchy must be rationalized. Most salary structures follow an automatic progression approach—a system of equalization. It is the simplest to implement. It requires a formula rather than managerial judgment. Differences between the best performers and the poorest have narrowed. Employee appraisals have skewed to the point at which most people are above average. In the process, organizations no longer meet their targets. Coombs and Gomez-Mejia, in "Cross-Functional Pay Strategies in High-Technology Firms" [4], raise the issues relative to the disparity in compensation between key strategic employee groups such as research and development and manufacturing and marketing. They argue that the compensation of the groups that are responsible for making the results of R&D commercially feasible (manufacturing and marketing) is negatively impacted by the high profile of R&D. They argue for a better balance and consistency in compensation as it applies to meeting the firm's strategies. Manufacturing professionals have been on the low rung of that compensation ladder for many years. The MBA programs did not and most do not even offer programs related to manufacturing. Why have advanced manufacturing systems (AMT) received such little attention, and why have they been a disappointment so often? The answers are simple: lack of competence in the new technologies and the new required management practices and an inability to manage human resources in a new environment.

Thompson and Scalpone, in "Managing the Human Resource in the Factory of the Future" [5], suggest five unique demands for implementing AMT:

- Concentrate on preparing people rather than eliminating them. As system complexity increases, more preparation is required.

- Work force knowledge and skills are more important in the AMT systems than in traditional manufacturing.

- Labor practices based upon command and control mentality do not fit AMT.

- Integration of functional disciplines is essential.

- Senior managers must set the performance objectives in a way that encourages functional cooperation.

The conclusions and recommendations of all the referenced authors seem logical and difficult to argue against. People are important, and the human resource department is important. The question that arises is how the human resource department can begin using its creative talent. Is creativity solely a prerogative of science and engineering? I doubt it. Developing a new appraisal form, defining a new salary structure, negotiating labor contracts, dealing with government bureaucracies, and attempting to influence senior managers in their choices for new appointments are routine human resource activities. Managers are increasingly being called on to think abstractly at multiple levels of analysis. They must also think in terms of systems that demand a new set of competencies. The new competencies will not be developed through the current approaches to continuing management education. Human resource professionals must put on their creative thinking caps and get beyond the routine administrative functions if their effort is going to make a difference in business performance.

I have identified one major effort that would involve a proactive response from a human resource department and the use of the latent creative talent. It would require the senior human resource executive to take a great deal of risk, but it would provide significant benefits: *Go back to the basics in dealing with people.* What do I mean? Over the past several decades, speed has entered our vocabulary. Learn a foreign language in 30 days; learn how to play a musical instrument in ten easy lessons; books are condensed; graduates in engineering are called engineers; graduates in the sciences are called scientists; and so on. The words "apprenticeship" and "trainee" no longer are appropriate titles.

It appears that few people want to acquire the fundamentals, use those fundamentals, learn from the application of the fundamentals, and build a solid foundation for a career. In many cases, they are discouraged from dealing in fundamentals. Consider the engineer: four or five years of education and no practical experience. In the past, this graduate would have spent some time on the drawing board—today on a CAD terminal plus the drawing board. An engineer's career may have started with tracing and progressed to various levels of design. The opportunities were to learn about some very fundamental issues in engineering: how to read engineering drawings, what it takes to complete not only a single drawing but a set of drawings for a new product or process, the interaction and interdependence of different working groups, and the many issues in becoming an engineer with a capital E. That type of learning by doing was a form of apprenticeship. After some years, the graduate finally became an engineer with depth and breadth of experience. Such an individual was a business asset. Depth and breadth were developed through progressive human resource policies, which must be uncovered and adapted to current conditions. Although the example is an engineer, it could be any professional or nonprofessional in the organization. Changing the system will involve controversy with the engineering graduate and management. Learning the basic requirements of any profession is essential for long-term performance. If nothing else, it pays off in an appreciation of those who do the more menial professional tasks.

Managing the Dynamics of the Business

From reading the business press, a conclusion that executives prefer change to stability and competition to no competition could be drawn. I have always questioned that conclusion. Certainly Kodak would prefer not to have Fuji Photo as a competitor. IBM would have preferred to live without the cloners of the IBM personal computer (PC). Detroit did not welcome competition from Toyota, Mercedes Benz, and other imports. Although change is generally considered to be something good because it forces people to think and to look to the future and seek that impossible dream, not many people can accept change and fewer will attempt to introduce it. Of course, change for the sake of change can be beneficial, but it can also be destructive. Continuous change, as some change gurus would advocate as a good without any proof, can be not only time-consuming but destructive as well. Change must be purposeful and must be balanced with periods of stability. The continuous reorganizations taking place in major corporations leave the participants with the feeling that no one is focusing on the real problems. The solutions are cosmetic, only to get a new face in the not to distant future.

It may be a truism, but business operates in a dynamic mode. The environment is continually changing in many different ways, some predictable and others not. However, what is certain is that business environment is not static. Unfortunately, too many business decisions assume a static environment and make decisions accordingly. If organizations are going to attempt to optimize cycle time, they must change the current attitudes.

Without complicating the issue with technical jargon, managing a business follows the three laws of motion described by Isaac Newton in the midseventeenth century. Newton's concepts of force, momentum, mass, and inertia lead to the description of the three laws of motion that are the foundations of classical dynamics.

- *Law of inertia.* Every body continues in its state of rest or of uniform motion in a straight line unless it is acted upon by an outside force.

- *Law of acceleration.* The change of motion of a body is proportional to the force acting upon it and takes place in a straight line along which the force is acting.

- *Law of equivalence of action and reaction.* For every action there is always an equal and opposite reaction, or the actions of two bodies on each other are always equal and opposite.

Those three laws of dynamics also apply to business change dynamics. In very simple terms: mass is a unified body (people, material, and so forth); inertia is nothing more than resistance to change (I'm not moving); force provides the action or the power against the inertia; momentum is the thrust in human actions (it's difficult to develop, more difficult to sustain, and once it's lost, it's exceedingly difficult to regain). The application of those laws must be reintroduced into the business unit processes. Executives complain about the inertia of the business system. How do executives eliminate the inertia or at least reduce it to the lowest acceptable limit? They use force. Force in this case means using all ten of the business resources to eliminate the inertia. What kind of force and how much force involves critical decisions? Just as in the law of inertia, one thing is certain: Unless some type of force is used, the inertia will remain. The law of acceleration tells the executive that change is proportional to force and that the change is going to take place along the line that the executive is pushing. In what direction does the CEO wish to move the mass with all of its inertia? The law of equivalence tells the CEO that there is going to be some opposition and that the opposition will in some way have to be neutralized through accommodation, negotiation, a different kind of force, or a redirection of the force.

Managing the business dynamics requires using all the various forces available to managers. Internal inertia can only lead to eventual dissolution of the enterprise. Weitzel and Johnson [6] discuss the gap between business equilibrium and performance as organizational decline continues. Although their study relates to W.T. Grant and Sears Roebuck, their interpretation of the continuum from business decline to dissolution probably applies to all industries. They describe five stages:

- *Blinded stage.*   Lack of understanding of the conditions that will affect the organization's long-term viability.

- *Inaction stage.*   Information that is available is not being used to develop corrective action at the appropriate levels.

- *Faulty action stage.*   Organization makes the wrong decisions, and there are no effective procedures for implementing the decisions.

- *Crisis stage.*   The organization no longer has sufficient resources or the effective mechanisms for improving the situation.

- *Dissolution stage.*   The organization closes and ceases to do business or is liquidated in an orderly manner.

Applying Newton's laws of motion to managing an organization may at first seem absurd, but the opposite is true. Weitzel and Johnson consider the dynamics of an organization. The next step is to realize that the organization does function in a changing environment and the principles of dynamic process control apply. Managers are really charged with the responsibility of keeping the infrastructure and the resources in stable equilibrium. Boynton and Victor, in "Beyond Flexibility: Building and Managing a Dynamically Stable Organization" [7], direct attention to organizations that manage contradictory requirements by becoming dynamically stable. They suggest that thinking only about today's processes, products, customers, and so on is insufficient. The key to strategic success (really operational success) is "being able to satisfy tomorrow's customer by developing process know-how and skills, that can adapt to changing and often unknown conditions." Dynamic stability in business is obtained in the same manner as dynamic stability in any manufacturing process:

- In-depth knowledge of the potential changes that could occur
- Know-how and human development capabilities that can respond to unpredictable events
- Creative use of data, information, and knowledge

Resolving the Information Paradox

Much has been written about the possibilities of using information for competitive advantage, but information as a single issue will not provide competitive advantage. Information is important, but by itself it accomplishes nothing. It is no more than the printed page of a book. If any change is to take place, some action must result from use of the information. However, the continuum that begins with data, progresses to information, and then is transformed into new knowledge must be clearly understood.

Porter and Millar, in "How Information Gives You Competitive Advantage" [8], attempt to make a case for the strategic significance of information technology: "The information revolution is sweeping through our economy. No company can escape its effects. Dramatic reductions in the cost of obtaining, processing, and transmitting information are changing the way we do business." They further state that the information revolution is affecting competition in three ways:

- It changes industry structure and, in so doing, alters the rules of competition.
- It creates competitive advantage by giving companies new ways to outperform their rivals.
- It spawns whole new businesses, often from within a company's existing operations.

Those statements were made by Porter and Millar in 1985. What happened to the revolution? If there was a revolution, it was a revolution that generated data and little information. It was a revolution that generated more internal junk mail than is processed through the U.S. Postal System. Structures have not been altered significantly because of the projected information revolution. Whether the impact it has for competitive advantage can be isolated from all other activities engaged in by any business is questionable. Although there is no doubt that the effective and efficient use of information can play a major role in business performance, there are no specific examples to show that information technology, as a single-issue approach to gaining a competitive position in the marketplace, provides any benefit. It is one factor in the technology-marketing equation for success. Use of information for competitive advantage must go beyond considerations for cost reduction and product differentiation. The critical decisions that managers face relate to *how much information is required.* It is not a matter of the cost; it is a matter of what information is required.

Cleveland, in *The Knowledge Executive* [9], makes clear distinctions among data, information, and knowledge.

- *Data* is the ore—the sum total of the undifferentiated observations—facts that are available to be organized by somebody at a given moment.

- *Information* is the result of somebody's applying the refiner's fire to the ore and selecting and organizing what is useful to somebody.

- *Knowledge,* which is derived from information, requires putting to use the available information in one's mind.

Cleveland adds that most knowledge is expertise in a field, a subject, a science, a system of values, a form of social organization and authority. He notes that wisdom crosses disciplinary barriers and integrates knowledge—information that is made superuseful by theory and that relates bits and fields of knowledge to each other, which in turn enables a person to use the knowledge to do something.

Cleveland's descriptions may help resolve the information paradox. Computers continue to provide large volumes of data with little information. The proof lies in the tons of computer printouts generated by business, government, and academia. Much of the so-called grunt work that is required to filter the information from the data continues to be performed by people. The computers have not been programmed to filter out the junk—the unimportant and the irrelevant. The computer can perform most effectively when it is used to extend human capabilities. That is true whether computers provide typical business financial results or scientific and engineering data. It applies to the control of complex computerized processes and the use of those processes in artificial intelligence and expert systems.

The current approaches to benchmarking provide an excellent example of the data—information—knowledge continuum. Benchmarking can be described simply as the *search for the best practices.* However, benchmarking is not something new. It has been practiced to different degrees of depth, breadth, and form. It involves two specific activities:

1. Comparison with the best in the business in a specific industry

2. Comparison of different functional operations not only among the competitors in a specific industry but with the best functional performers in any industry

Benchmarking can provide significant advantages. Point 1 is closely allied with the various long-range and strategic planning processes and is then referred to as competitive intelligence if it goes beyond product and sales and profit considerations. As a matter of fact, it is impossible to chart the future of a business without that information. Comparison with the best in the industry involves knowing the organization, the customers, the industry, the suppliers, and the competitors and having some sense of the social, political, and economic climate in which the business operates.

Point 2, benchmarking the various functions in an organization against the best, regardless of industry, is a relatively new concept not often practiced. It focuses attention outside the specific industry and attempts to determine who are the best in certain specific functions. That includes such areas as customer service, time to market, new-product introductions, purchasing, fleet management, and technology—the list would include every activity in the organization. As an example, what can the automobile manufacturers learn from fast-food chains; what can any manufacturing organization learn about order processing from the financial industry; what can a software developer learn from R&D operations; what can the discrete parts manufacturers learn from the continuous processing industries. If benchmarking is to be useful, it must go beyond the data collection stage. The key decision that must be made before any data are accumulated is to focus the question. Attempting to compile all types of data on the industry and the major competitors will only yield a great deal of data. The objective must be focused. The data must be transformed into usable information and, through application, become knowledge that will help guide the business directions.

Dealing with Discontinuities

Drucker, in *The Age of Discontinuity* [10], introduced the world to the topic of discontinuities in technology, world economics, and the political matrix of social and economic life. The study of discontinuities from a mathematical perspective is not new, but the application to the issues faced by a changing society forces a review of the changes that have taken place and those that are in progress. Drucker's approach does not provide the exactness of the mathematicians, but it does provide a conceptual framework for looking at the issues facing society. Drucker looked at the macro societal system, but the discontinuities related to technology and knowledge have the greatest impact on cycle time. Foster, in *Innovation: The Attackers Advantage* [11], describes technological discontinuity as the "periods of change from one group of products or processes to another," but that description is not applicable to technological discontinuities only. Gaynor, in *Achieving the Competitive Edge through Integrated Technology* [12], extended the concept in a generic sense as *the periods of change from one behavior to another.* It applies to all discontinuities including those in the marketplace, in information processing, and in the management processes.

Knowledge is expertise, and it derives from the application of information to the resolution of problems. It follows the same S-curve principles as those in technologies and management. As an example, a child's learning processes begin at essentially zero and follow the typical S-curve. Once a child learns the numbers from 1 to 10, new horizons of understanding appear. Until then, little progress can be made in understanding the use or importance of numbers. That is true of the letters of the alphabet also. Knowledge begins with learning, gaining information, and then doing something with the information. The two are interrelated. Knowledge is gained by learning a little, doing a little, learning a little more, doing a little more, and so forth.

Discontinuities are not limited to technology and knowledge. They are present in marketing, human relations, finances, and management. Every aspect of business life could be a victim. Avoiding the negative impact of discontinuities requires looking at the future rather than taking satisfaction with the past. Management discontinuities exist today in many blue-chip corporations. The total emphasis on reducing middle management and general downsizing stems from a lack of recognizing the impact of global competition and impending economic changes. A clear discontinuity exists between short-term and long-term thinking. Short-term thinking in all aspects of management reaches an endpoint, and then another short-term effort must begin and the process continues. A single issue is followed by another with a period of discontinuity before another begins. When management made a decision to downsize, a discontinuity occurred. Senior-level managers may not admit it, but they made the decision to downsize without providing for filling the new gap.

Discontinuities in management add cost. Foster has documented the costs of technological discontinuities, and managers are becoming more sensitive to their effect on business performance, but the cost of managerial discontinuities is seldom considered. Management has the responsibility to balance stability and change. It also has a responsibility to look to the future. To a greater or lesser degree, every change creates some conflict. The level of conflict determines the degree of discontinuity. The recent actions by IBM and General Motors have created discontinuities. Both companies were riding the S-curve and had reached the point at which added investment of resources was not providing any significant benefit. The economy certainly has affected their performance, but the economy is part of the management-related S-curve analysis. This type of major discontinuity could have been prevented or at least minimized if the respective senior managements had extended their thinking beyond determining the future by extrapolating the past. A responsive management system will not eliminate risk from the business equation, but it can minimize the impact.

Measuring Performance

Measuring performance is a topic most managers prefer to delay to the next meeting, and the subject is seldom seen on the agenda. The statement is often made that executives and managers are numbers-oriented. If that statement were true, CEOs would not be attempting to rationalize their current lack of performance in an economic downturn. The numbers show that many organizations had and continue to have more people than are required to operate the business according to the three E's. The numbers also raise issues of effectiveness and efficiency. The excess complement of people has been there for years, but management did not see it. The inefficiency was there, and the managers discounted its impact on future performance. They looked at one line and even managed to convince Wall Street that they were lean and mean.

Cameron, Freeman, and Mishra [13] provide some interesting statistics relative to performance:

- Private sector productivity growth slowed from 3.3 percent per year between 1948 and 1965 to 0.1 percent today for the entire economy.

- Productivity for nonfarm business declined by 0.3 percent. Between 1978 and 1986 the number of production workers declined by six percent while real output rose fifteen percent. That represents a gain of 21 percent in blue-collar productivity, or a 2.4 percent annual growth. During the same period, U.S. manufacturing organizations expanded the number of white-collar nonproduction workers 21 percent, representing a 6 percent decrease in productivity.

Their exactness may be questioned, but such statistics cannot be ignored. The results can be attributed only to the malpractices of managing. One of those malpractices is the reluctance to measure performance, beginning with the CEO. There is no question that factory productivity can be measured accurately, at least in terms of averages. Individual performance measures become clouded by justification for lack of materials, machine downtime, and many other factors. Even in the case of machine operators, individual productivity can be measured if some assumptions are made and are consistently followed throughout the measuring period. The productivity of salespeople also can be measured if the appropriate limitations are factored into the equation. The performance or the individual contributions of the professionals—accountants, engineers, scientists, marketers, and attorneys—although inherently less precise, must be measured somehow. The closer the measurement approaches the undefined or the nonprogrammable work effort, the more difficult the task becomes. That is true of all executive and managerial levels.

Failure to measure performance against some predetermined criteria is responsible for much of the current organizational downsizing. The significant increase in white-collar and professional staff is certainly responsible for lower productivity.

The measurement process

Measurement can be accomplished only if objectives are clearly described. Without objectives, there is no standard against which to measure progress. Measurement does not imply developing numerical indices to the second decimal place and then rank-ordering. The measurements will be essentially qualitative and based on judgment. Qualitative measurement often provides more information than quantitative measurement. As an example, when corporations use quantitative figures such as increased sales per employee or dollars spent in research, just what do the figures imply? Sales per employee could have risen for any number of reasons that have nothing to do with improved employee performance. Spending more on research does not necessarily guarantee more new products. At best, all measures of performance bear some element of the qualitative.

One very simple approach to measurement of professional performance is a yes or no response. Assuming that objectives for each individual and the department have been described, the approach is to count the number of yes and no answers. There are no in-between responses and no excuses for nonperformance. Objectives are not limited to the performance of the immediate work at hand; they could also relate to additional education, interdepartmental communication, resolving conflicts, performing studies, or working on related task forces and committees. Objectives relate to all individual's and the department's activities. The challenge is to break them down into measurable yes and no answers. But to use a system of that type requires a change in mentality by management. The batting average is not going to be 100 percent. Furthermore, since the measurement of individual performance depends on the contribution of at least the immediate manager, the manager must take an interest in the individual's effort. As the process continues, managers and the participants eventually begin to understand the necessity for the up-front work in establishing measurable objectives. Objectives are not wishes. As time passes, groups can accomplish most of their objectives because they begin to understand what has prevented them from meeting their objectives in the past. They understand the objective and know if the ten resources and seven elements of the infrastructure support the objective.

Krogh [14] provides a quantitative approach to measuring laboratory productivity and quality. He considers a measure of return on laboratory funds employed (ROFE).

$$\text{ROFE (\%)} = \frac{\text{new product sales}}{\text{expenses}} \times \text{new product profitability (\%)}$$

This approach provides a numerical value for a laboratory but not individual performance. New product sales may or may not be affected by laboratory performance. The effectiveness of marketing and sales affects sales revenue as much as laboratory product quality. New product profitability may have little to do with laboratory expenses. As a macro measurement, it does allow comparison of one laboratory against another.

Operating a research or development department as a business unit also provides opportunities for measuring performance. Basically, the group functions as an independent business and operates along the lines of independent research laboratories, design engineering consultants, and architectural firms. Every activity relates to a project, and participants charge their time to specific projects. There is no need to become obsessed with the detailed numbers, plus or minus 5 percent is more than adequate. The system does influence the discipline of an organization in meeting objectives, schedules, and costs.

No precise techniques are available for every situation, and organizations must experiment with approaches that provide a comparative measure of performance that satisfies their needs. Projects are not completed at the end of the calendar or fiscal year. They occur throughout the year. Assumptions about performance will have to be made. There are other considerations: size of the particular group, staffing of the group with the required experience and skill, the organizational structure, the value added by the group to the business, and the time spent on various classes of activities (the group may designate certain activities as primary or secondary). The simplest and most effective method used by the author is a yes or no response to meeting the project requirements of performance, schedule, and cost according to a predetermined plan. Each element must be broken down in sufficient detail that it can be isolated as a measurable performance. Thor, in "Performance Measurement in a Research Organization" [15], suggests the following discrete activities

- Department cost
- Productivity described by milestones, use of intellectual resources, and so on
- Market success
- Optimum quantity of work in process
- On-time performance
- Cycle time—elimination of waiting periods, measurement of start-to-finish times, and so on
- Documentation
- Creativity and innovation
- High science described as the availability of individuals or equipment associated with world-class state-of-the-art practice—this should be changed to leading edge
- Safety

Exactness is not the objective. The measuring system is not the objective. Measuring is a tool for helping achieve the objectives.

Expecting Innovation from all Functions

Management has no qualms about asking for creativity and innovation from the technology-related functions. Every year, senior management calls on manufacturing to reduce costs and improve quality. The concept of expecting improvement must be extended to the total organization. Every function should be expected to provide its share of creative input and innovation. Each business function has creative resources that have been dormant. The need for creativity and innovation in human resources has been mentioned. Business conditions demand that all functions deliver their fair share of creative thought and follow-up with the determination expected of the technology functions. The results would reveal that a new order of cooperation will occur.

1. Japanese enter the market: low price.
2. Corporate strategy: match price and retain market share; lose money if necessary.
3. 1983. Portables adopts a new program.
4. 1987. Sales have increased; cycle time is reduced from 25 weeks to 7 days; inventory levels are down by 80 percent; work-in-process is reduced to 125 units from 1500; floor space is reduced by 50 percent; the number of vendors is reduced from 1500 to 200; product and service quality are improved; more than 70 percent of sales are delivered within 2 days; and market share and profitability are excellent.

Figure 10.2 Chronology of events associated with changes occurring at the Portables Division of Tektronix.

One may question the need for an accounting department to be creative. Business publications often remind us of the negative effects of creative accounting. Many current accounting practices are restrictive and obsolete because they do not present the situation as it really is. Good product programs and businesses are often discontinued because of the manner in which costs are allocated. Turney and Anderson, in "Accounting for Continuous Improvement" [16], relate the changes that occur when accounting becomes a facilitator rather than a watchdog. They relate the history of the Portables Division of Tektronix. The details are shown in Fig. 10.3.

With that effort, Portables set the standard for price and performance, and it did so by continuous improvement in manufacturing, engineering, and marketing. Innovations in accounting systems and

1. Removed accounting systems that no longer provided useful information.
2. Collected information based on the output of the production line.
3. Substituted exception reporting for detailed reporting of direct labor time.
4. Focused not only on cost but also on quality and delivery.
5. Eliminated waste and scrap reports when level dropped to such a point that tracking was no longer required.
6. Eliminated extensive inventory tracking system after JIT was installed.
7. Allowed standard, labor, material, and overhead costs to be charged to completed units when expediting of orders was no longer required.
8. Managed day-to-day production with visual controls.
9. Targeted manufacturing overheads for continuous improvement: purchasing, inventory control, expediting, and so on.
10. Reduced reporting procedures. Direct labor represented 3 to 7 percent of manufacturing cost; recording and analysis time was greater.

Figure 10.3 Actions taken by accountants in the Portables Division of Tektronix. The actions were taken because of the involvement of the accountants in the project process.

approaches were a prerequisite to Portable's successful turnaround. The team that began the turnaround included accounting as a participant. A dual reporting system made possible the tracking of management and financial costs. Accounting innovation requires leadership and people who are willing to experiment. Instead of providing formalized reports, the accountants spent their time providing useful information to the team. Most accounting systems collect data. As has been stated previously, they are the historians. In this case, the accountants were participants, and they helped to introduce meaningful financial information through the actions listed in Fig. 10.3.

The Tektronix example demonstrates how creativity, coupled with integration into the business, provided an accounting department with the opportunity to make a significant contribution. Similar opportunities are available in every staff function. It takes a creative management to tap the creativity of its people resources.

Gaining Advantage from Organizational Learning

Industry spends vast amounts of money on education and training. Even though American industry lags significantly behind Japan and Europe in investing in education as a percent of sales, the total represents an expenditure of a significant amount of resources without any measurable payback.

Combining all the different types of learning into one category called *learning* does not do justice to such a major component of business success, primarily in the utilization of its resources. Learning must be considered as a multicomponent element of the business. It includes: individual learning, project, team, or group learning, and business-level learning. Project, team, or group learning is just what the name implies: It is the learning that takes place in a group as the group goes about meeting its objectives. As groups become involved in activities, they learn from one another. There is a spillover from one individual to others in the group. The synergy that develops allows the accomplishment of something beyond the capacity of any one individual. It is the sum of the learning that takes place within a group—the collective learning. Any group, even with the highest level of competence, cannot achieve the objectives if collective learning does not occur within the course of the group's activities. That collective learning becomes part of the business's intellectual resources.

Business level learning embodies the same elements as those attributed to the group. In addition, it includes the collective learning of all the groups within the business. It is the sum of all the learning that has taken place in the past or is taking place at the present time. This learning can be a powerful tool to improve performance if organizations find a way to capture that learning and use it. It is an intellectual resource, yet organizations disregard former learning experiences. Disregarding the past learning extends cycle time. Much of the difficulty arises from the fact that managers continue to ignore their roles as teachers. Allowing everyone to start at the bottom of that learning curve is a misuse of human resources. Organizational collective learning must be exploited to the fullest. It is not only a resource, it is also a business asset. It is the collective learning that allows executives who fully understand their businesses to manage them according to the three E's. The concept of collective learning must be driven down through all levels of the organization.

Learning by doing has been an approach followed by many organizations. The concept is very simple: learning takes place as individuals engage in activities and attempt to solve problems. It is doubtful that the learning-by-doing approach, by itself, is viable today—or if it ever was. The fact remains that a little book learning plus some coaching and teaching provides a much faster approach to advancing people up the learning curve so they can begin making a contribution to the business by participating in related activities. Udayagiri [17] shows that learning by doing can be extended to new activities, new products, and future generations of current products. He states that the knowledge gained from learning by doing has been viewed as productivity enhancing. He argues that it can be a source of innovation. Technological learning is dependent not only on the knowledge generated from learning by doing but also on the organization's capacity to absorb the knowledge. He describes technological learning as the outcome of knowledge spillover and the absorptive capacity, and he says that the knowledge spillover can be useful in other products, future generations of products, and other functional activities in the firm. Udayagiri's conclusions appear to be self-evident, but organizations continue to disregard the learning of the past, which in many cases has been a very costly exercise.

Managing the Process While Focusing on Substance

The profit motive is an essential ingredient of a market-driven economy. Businesses must certainly manage the infrastructure and the resources in such a way that some residue remains after all the expenses are accounted for. The residue is a result of a process. Quality, productivity, and profits are the measures of performance of the management process. U.S. managers focus on the results; Japanese managers emphasize the process and the continuous improvement of the process.

That attitude is the one major factor that differentiates U.S. managers from their Japanese counterparts. Neither approach, by itself, is adequate. The expected results must be kept at the forefront, and the process must be tuned to achieve the desired results.

Most businesses operate according to the sitcom model, which represents the single-issue management approach that governs the culture of U.S. industry. A more appropriate model might be managing a business as continuous drama with no interruptions. The plot rolls on from day to day with issues being resolved and new ones arising. The participants, because of their specific talents and the roles they play, enter and exit and reenter as the occasion demands. The different characters bring with them their creativity, personalities, perspectives, and all their idiosyncrasies, which influence and are influenced and are subsequently directed and molded into a configuration that enhances the process of creating business advantage through continuous change.

Managers must begin thinking about their businesses as continuous flow operations. The intermittent starts and stops cost businesses billions of dollars annually. Achieving the benefits from system cycle time management (SCTM) requires that managers begin to respond to new business challenges with the tools that are required to manage the process. All the principles derive from the fundamentals of managing; there are no mysteries. In that sense, optimizing cycle time is a *result* of good management practice throughout the organization rather than the addition of a new program.

Managers emphasize results to the exclusion of the process. New programs are introduced and the entire focus is on results. How those results are achieved seldom receives consideration. Most of the activities related to quality, strategic planning, concurrent engineering, and so forth are process-oriented, but that process is often devoid of substance. The assumption is made that going through the motions will yield the expected results. The CEOs who emphasized strategic planning in preference to strategic thinking and strategy thought that the process would lead to dominating the market.

Although the process is associated with achieving results, it must be managed. It involves writing the human algorithm by which it will be controlled. "Control" does not mean rigid rules and procedures to enforce compliance. An algorithm used in managing a chemical process with many variables provides only the amount of control required. Some parameters may need very accurate measurement and control; others will be given a much greater latitude. Some raw materials may require very rigid specifications and close control of their introduction to the process; others may have much wider latitudes. The development of the algorithm takes those differences into consideration. The results are determined by the algorithm that manages the process, and the process of managing can be guided by the same fundamentals.

Winning in sports is what counts, and that is true of a business. It *does* make a difference how the win is accomplished. Teams can win by playing as though they were suffering from klutz fever, or they can win by demonstrating all their knowledge and skill by doing it like a bunch of pros. Winning in sports through some default, through luck or misfortune in the last seconds of the game, leaves the spectators with a sense of emotional ambivalence. How individuals or teams win depends on a process that includes substance, content, and meaning. It is not just going through the motions. The process requires thinking and creativity.

Implementing SCTM is a process that requires adding the substance, the content, and the meaning. From a macro perspective, the process is simple: introduce SCTM as a business integrating activity; support it with a viable infrastructure; establish the objective; spread it throughout the organization; allocate the resources; and work, work, work, and work until the system delivers the expected results—continue the process to reach new levels of performance. The target is moving, and what was acceptable yesterday may not be acceptable tomorrow.

Landen and Landen, in "Understanding—and Improving Organizations through Cybernetics" [18], confirms my observation that process and outcome (substance) are one. They are inseparable because process, to be useful, must shape the outcome and be influenced by it. Unless that condition is met, process is strictly a means for stumbling through in some mechanistic way toward a decision. Landen and Landen also confirm the basic tenet of this book that single-issue programs do not work. The behaviors that produce the discontinuities and limit the achievement of the performance expectations are embedded in the organization's philosophy of operation. Performance is affected by attitudes exhibited by management and the other participants. Unless attitudes of the leadership and the participants are time- and event-synchronized, management effort to change the system works against that immovable object discussed in the section on dynamics.

Summary

System cycle time management is a major business issue. It requires a responsive management system that includes all the elements of the infrastructure, the availability of adequate resources, and attention to operational management issues:

- *Human resources.* Putting those statements of purpose in relation to employees into practice.

- *Functional linkages.* Thinking in terms of the business system rather than the parts.

- *Dynamics of the business environment.* The three laws of motion that are the foundation of classical dynamics apply to managing: Nothing will change without the application of some outside force.

- *Paradox of information.* Clear distinctions must be made among data, information, and knowledge. Information is vital for survival, but organizations must learn how to turn it into useful knowledge.

- *Discontinuities.* Discontinuities appear in every business function, and not just those related to technology. They cannot be avoided, but they can be accommodated.

- *Measurement of performance.* Measuring individual, group, and business unit performance is a must. Qualitative measurements, if performed with consistency, provide as much information as—if not more information than—the quantitative measurements.

- *Organizational learning.* Individual learning must be leveraged and transformed into organizational learning. Knowledge of one individual must spill over to others; it is too important to be limited by the capability of any one individual.

- *Creativity and innovation.* Raise the expectations, and insist that all functions and individuals contribute their share to the development of the business. It takes more than the creativity of engineers and scientists to build a business.

- *Process with substance.* Process is essential, but it must allow for depth and breadth analysis and synthesis. There is no long-lasting quick fix.

References

1. G. M. Bounds and L. A. Pace, "Human Resource Management for Competitive Capability," in M. J. Stahl and G. M. Bounds (eds.), *Competing Globally through Customer Value,* Quorum Books, New York, 1991, pp. 648–707.
2. I. Meshoulam and L. Baird, "Proactive Human Resource Management," *Human Resource Manage.,* vol. 26, no. 4, Winter 1987, pp. 483–502.
3. L. Baird and I. Meshoulam, "Managing Two Fits of Strategic Human Resource Management," *Acad. of Manage. Rev.,* vol. 13, no. 1, 1988, pp. 116–128.
4. G. Coombs, Jr. and L. R. Gomez-Mejia, "Cross-Functional Pay Strategies in High-Technology Firms," *Compensation and Benefits Rev.,* September-October 1991, pp. 40–48.
5. H. Thompson and R. Scalpone, "Managing the Human Resource in the Factory of the Future," *Human Resource Manage.,* vol. 5, 1985, pp. 221–230.
6. W. Weitzel and E. Johnson, "Reversing the Downward Spiral: Lessons from W. T. Grant and Sears Roebuck," *Academy of Management Executive,* vol. 5, no. 3, 1991, pp.7–21.
7. A. C. Boynton and B. Victor, "Beyond Flexibility: Building and Managing the Dynamically Stable Organization," *California Manage. Rev.,* vol. 34, no. 1, Fall 1991, pp. 53–66.
8. M. E. Porter and V. E. Millar, "How Information Gives You Competitive Advantage," *Harvard Bus. Rev.,* July-August 1985, pp. 149–160.

9. H. Cleveland, *The Knowledge Executive,* Truman Talley Books/E. P. Dutton, New York, 1985, pp. 22–25.
10. P. F. Drucker, *The Age of Discontinuity,* Harper & Row, New York, 1968.
11. R. N. Foster, *Innovation: The Attacker's Advantage,* Summit Books, New York, 1986.
12. G. H. Gaynor, *Achieving the Competitive Edge through Integrated Technology Management,* McGraw-Hill, New York, 1991, pp. 187–207.
13. K. S. Cameron, S. J. Freeman, and A. Mishra, "Best Practices in White-Collar Downsizing: Managing Contradictions," *Academy of Management Executive,* vol. 5, no. 3, pp. 57–73.
14. L. C. Krogh, "Measuring and Improving Laboratory Productivity/Quality," *Research Management* now *Research · Technology Management,* November-December 1987, pp. 22–24.
15. C. G. Thor, "Performance Measurement in a Research Organization," *National Productivity Rev.,* Autumn 1991, pp. 499–507.
16. P. B. B. Turney and B. Anderson, "Accounting for Continuous Improvement," *Sloan Manage. Rev.,* vol. 30, no. 2, Winter 1989, pp. 37–47.
17. N. D. Udayagiri, "The Strategic Implications of Technological Learning: A Framework for Analysis," *Academy of Management Annual Meeting,* Miami, August 1991.
18. D. L. Landen and G. A. Landen, "Understanding—and Improving—Organizations through Cybernetics," *National Productivity Rev.,* Winter 1990/91, pp. 5–17.

Costs of Accelerating
Cycle Time

Speed to market, but what speed? The proponents of fast cycle time disregard the costs that may be associated with it. Being in a hurry to reach an objective can consume a disproportionate amount of business resources. Speed to market can be compared to speed required to reach a particular destination. It is not a question of speed per se; it is a question of the speed that is right for the physical and environmental obstacles that lie in the way of reaching the destination. Speed in the automobile can kill; speed to market has the potential of leading organizations to the bankruptcy courts. The issues related to the costs of accelerating cycle time include:

- Using the fundamentals

- Doing things fast—pushing the process

- Challenging current practice

- Justifying the selected alternative

Using the Fundamentals

Simultaneous engineering has received a significant amount of publicity, but many of the claims for its usefulness are unsubstantiated. What does a statement such as, "Hewlett-Packard Co. (HP) decided to improve the quality of its products by 1000 percent in five years—and succeeded" [1] really mean? The improvement is claimed to have occurred because of the use of multifunctional and multidisciplinary teams in a concurrent (simultaneous) engineering setting. It is difficult to comprehend that HP's quality could have been so bad. The premise that speed to market as a single issue in today's competitive environment will make a difference is unfounded. What is important in introducing new concepts, whether related to products or processes, is that the concepts be introduced at the right time and with the optimal use of resources relative to the estimated benefits. Although the emphasis in this book has been on technology, products, and processes, the same conditions apply to introducing any new activity. There is no conceivable reason why a program originating in a staff function (human resources, financial, and so on) should be treated differently than a technology or product program. Personal experience indicates that such programs should receive greater scrutiny because of their downstream impact on cycle time. Optimizing the cycle time for introduction of any concept or activity will most likely add cost, at least in the near short term, and that cost must be measured against some future attainable objectives. Crosby [2] argues that quality is free, but it also requires an up-front investment so that at some time in the future it will not only be free but contribute to the business. Cost must always be evaluated against the potential benefits to the organization, but not only in terms of bottom-line results. The benefits must be viewed from a much broader perspective than the immediate activity in question. They must be considered in relation to the short- and long-term impacts on business performance. They must include the quantitative and the qualitative aspects of the situation.

A list of product or process-related projects in any organization includes projects of different scope, complexity, and importance. There may be many small projects that are managed by a group that has responsibility for managing them in a routine manner. They could relate to research, development, manufacturing, or marketing. On small or routine projects it is not necessary to consider each project individually and determine if the cycle time can be reduced. The emphasis must be placed on how the small projects, as a group, can be completed without negatively impacting the business objectives. Major business projects that affect growth must be considered individually, however. The major projects affect cycle time to a far greater extent.

Christoph-Friedrich von Braun [3] provides the following statistics on a company with average age of product that has been decreasing over time. In 1975, 40 percent of the sales represented products that had been introduced less than 6 years before, 33 percent were products introduced within the preceding 6 to 10 years, and 27 percent were products more than 10 years old. By 1986, sales of products introduced less than 6 years previously had increased to 56 percent, the 6- to 10-year group was down to 29 percent, and the more than 10 group was at 15 percent.

Such figures must be viewed very cautiously. They show progress, but the interpretation depends on how an organization describes new products. The percentages cited were most likely not of new-to-the-market products. The figures may console corporate executives, but they conceal reality. Percentages have no meaning unless the base is known: Percentages of what? Reductions in revenue and profits allow executives to find creative ways to rationalize nonperformance. Part of that nonperformance is due to a lack of new-to-the-market products. The increases in what are referred to as new products are most likely improvements, modifications, or replacement products that do not stimulate any significant growth unless the market is expanding. They serve the same market and perform essentially the same functions. There is no question about the importance of introducing product improvements and replacements, but they are generally shorter-term activities that lack the required changes to stimulate economic growth.

"Time is money" is an accepted truism; investments are expected to produce certain predetermined benefits. Figure 11.1 shows the investment over 10 periods in time for introducing a new concept, technology, product, and so forth. The investment is minimal at the start, reaches some maximum level, and tapers off. There is no typical curve that applies to every situation. Figure 11.1 shows the investment of all resources that can be translated into monetary terms; it accounts only indirectly for the nonquantifiable resources. When the latter are taken into account, the cumulative investment takes the shape of the S-curve in Fig. 11.2. The S-curve also is based on 10 periods. Figure 11.3 shows some different possibilities for evaluating the time-cost tradeoffs. To simplify the explanation of Fig. 11.3, assume that the investments per period of time shown in Fig. 11.1 have been averaged and are constant during the 10-period cycle; that is, the same amount will be invested in each period. The total investment is now represented by the area *ABCD,* which represents the total investment accomplished in 10 periods. Now make the assumption that an organization wishes to decrease

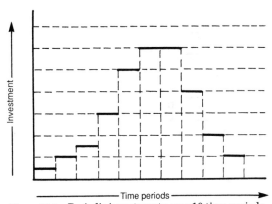

Figure 11.1   Periodic investments over 10 time periods.

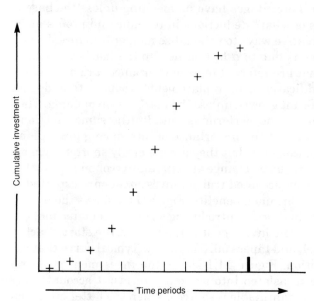

Figure 11.2 Cumulative investment over 10 periods of time as shown in Fig. 11.1.

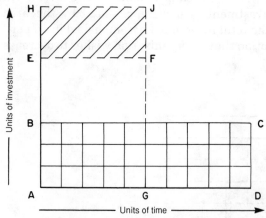

Figure 11.3 Cost possibilities associated with accelerating cycle time.

the cycle time by 50 percent, which reduces the time from 10 periods to 5. Assuming that no additional investment of resources was required, the area *AEFG* represents the total investment. If that were possible, an argument could be made to use the shorter cycle time. The decision does not pose any problems. Now consider the situation in which additional investments are required to achieve a reduction in cycle time. The area *AHJG* now represents the new total investment.

The area *EHJF* represents the added investment. The greater the deviation in total investment, the more difficult it becomes to justify shortening the cycle time. However, there are two parts to the justification: (1) Justification based solely on financial considerations and (2) justification based on the necessity for remaining in business. The first is short-term-oriented, and the second determines the long-term growth. The decision becomes more complex, and some justification must be developed for the added investment. Accelerating cycle time must take into account the total system cost and not just the costs associated with R&D. It is not a matter of justifying an investment in a particular product development project; it is a matter of justifying the investment on the basis of providing specific benefits to the business. The decision to shorten the cycle time by some percentage must be weighed against the benefits gained in the marketplace. The analysis must address not only the opportunity and financial aspects of a single project but the impact on the current line of products it may support, a new line of products that it may create, and finally its impact on the total business. That analysis must take into account the areas of the business that can be leveraged in the new project.

Figure 11.3 illustrates what can be considered as the two extreme ends of the cycle time continuum. There is little doubt that shortening cycle time will involve an additional initial investment. Assuming that the investment illustrated by the area *ABCD* was realistic and took into consideration all the required resources, compressing the cycle time will require additional resources. This is not a question of making an estimate of the required resources and then deciding what can be eliminated or how shortcuts can be taken to decrease the cycle time. Those reductions were taken into account when the area *ABCD* was calculated.

Accelerating cycle time involves trade-offs between cost and time. Deciding on an alternative depends to a great extent on the current effectiveness of an organization. If the system is inefficient, the costs may be minimal: those associated with education and creating a new environment. Significant added resources may not be required. However, an organization that is already effective and efficient may require greater resources to gain a further time advantage. Steady progress over the years has provided the organization with extensive benefits, but now additional investments yield marginal benefits. The current situation could be considered a discontinuity. The organization must do something new to achieve continued progress. This is another example of the discontinuities associated with managing a technology, product line, or business up the S-curve.

Research associated with the costs and benefits of accelerating cycle time is inconclusive. There is insufficient evidence to attempt to provide a heuristic approach that would have general application. Some of the research is contradictory and all of it is anecdotal. That is to be expected, since the research has not included a sufficient number of organizations in any one industry from which to generalize a conclusion. At the same time, there is no need to ignore the findings. Contractor and Narayanan, in "Technology Development in the Multinational Firm: A Framework for Planning and Strategy" [4], report on the work of Mansfield [5] regarding the time-cost trade-off. Mansfield found that the Japanese operate at a point at which costs would increase by 9 percent for a reduction of 1 percent in cycle time, whereas U.S. industry operates at about a 4 percent increase for a 1 percent decrease. The authors conclude that the Japanese apparently place a greater emphasis on first to market, but there is no evidence to support this conclusion. One of the major and unresolved difficulties lies in the comparison of figures. The authors do note that the greater percentage allocated by the Japanese is on a lower base, since the Japanese allocate a much greater part of their development cost to tooling and manufacturing facilities: 44 percent against 23 percent for the United States. They also cite the benefits received by the Japanese from the use of external technical resources. As an example, the value added–to–sales ratio for Toyota (Eckard, 1984 [6]) was about 15 to 20 percent compared to 48 percent for General Motors. There is no doubt that the Japanese bring products to market in a shorter period of time, but the reasons differ from one organization to another. If there are common threads, they involve doing the up-front work diligently, emphasizing process from the beginning stages of a new product, and follow-through from research to manufacturing.

Contractor and Narayanan also state that the trade-off between project duration and project cost is depicted as a convex curve. The greater the time compression the greater the cost according to some geometric multiplier. That may not necessarily be the case. There is no single or simple explanation for the shape of the curve. The shape depends on the type of product, the complexity of the technology (pushing the frontiers will be quite different from working with proven technologies), the number of technologies, the number and type of people involved, and all the other parameters that affect the time-cost relationship. There is no simple formula that can be applied across industries or across the differences that span the process industries and the piece part or equipment manufacturers.

Cost increases associated with reducing cycle time do not arise only in R&D. They include manufacturing as well as marketing and the supporting functions of administration. Isolating the cost increases becomes important because the magnitude depends on where the costs are generated. Doubling a development staff may not seriously affect the total project cost. As a matter of fact, doubling the people resources in any activity may be considered an incidental cost. In the long term, such an action may significantly reduce total cost because of the quality of the up-front work. Vesey, in "The New Competitors: They Think in Terms of Speed-to-Market" [7], makes the point that product design may represent an *apparent* influence on cycle time of only 20 percent but wield a far greater *real* influence. That "real" influence probably exceeds 50 percent, and it is due not to the intrinsic nature of the design, but to the impact and its effect on the downstream activities. The design determines the raw material used, the production equipment required, and the difficulties that will be encountered in production. Although Vesey mentions the impact of the differences between *apparent* and *real* influences on the design-engineering-manufacturing activity, it is necessary to recognize that the differences (apparent as contrasted to real) exist in all the business functions. Any business function at any one time can influence cycle time far beyond its particular sphere of activity and responsibility.

Vesey also provides the results of a study by McKinsey & Co. that shows the impact on performance if a product misses its target date to market by 6 months.

- Gross profit potential is reduced by 33 percent.

- Improving time-to-market by one month will improve profits by 12 percent.

- For revenues of $25 million, annual gross profit increases $400,000.

There are some underlying assumptions that must be considered. Not every project that reaches the marketplace 6 months in advance of its scheduled time will provide better results. As has been stressed throughout this book, cycle time is not the only issue. Total time for accomplishment and timing must be integrated with the considerations of cycle time.

Graves, in "Why Costs Increase When Projects Accelerate" [8], considers some of the issues related to R&D. Those he cites are generally accepted by managers associated with product R&D. As has been noted, the desire to combine research and development into the multifaceted function called R&D does not take into account the differences in contribution to cost by each. Graves suggests the following reasons for increased cost due to acceleration:

1. R&D is a heuristic or experimental process, and each step depends on the results of the preceding steps. Attempting simultaneous activities with high degrees of uncertainty and inadequate information results in mistakes and the additional use of resources.

2. The law of diminishing returns dictates that at some point in time adding people resources provides only a marginal benefit. The need to bring the new participants up on the learning curve to the point at which they can make a contribution absorbs the time of the experienced team members.

3. The probabilistic nature of some steps in R&D also adds cost when accelerated. If a single activity can be approached in 10 different ways, each activity has the same probability of success and the same cost, and all are independent of one another, the minimum cost strategy would be to attempt each in sequence and stop the activity when a successful solution is found. All 10 could be tried concurrently, but cost would be added.

4. If an R&D project is viewed as a network of activities, the objective would be to find the critical path through the network. The next approach is to find a way of compressing the critical path. The manager can then choose to accelerate the easier activities followed by the least expensive activities. If more compression were required, more costly choices would be exercised. The process could continue until an overlap of activities would require concurrent acceleration.

Graves considers the above factors to be the four that increase cost when projects are accelerated. They are obviously debatable, and they may or may not apply depending on the industry, the business, and the many factors that determine an organization's approach to research and development. The reason for making the distinction between research and development may now be more apparent. Research may be a heuristic process, but development is not, or at least it is not in the same sense or to the same extent. The typical development project deals with known technologies: technologies that have been applied and proved. There may be extensions of some technologies that must be advanced by further research, but such research efforts are usually of short duration. However, it is not necessary to restrict research activities to a set of constraints that forces the step-by-step approach, at least not in research related to new products. Research rigor may demand the step-by-step approach, but research related to new products can focus attention on the advantages of taking the giant step when it is appropriate and feasible—in effect, doing the last experiment first.

Although all the facts may not be known in the giant step approach, sufficient information permits applying it and developing new knowledge. Waiting for all the evidence may only delay product introduction and fail to take advantage of what can be learned by early introduction to the marketplace.

Overstaffing, with the intention of accelerating cycle time, may result in only a marginal contribution to the project. The law of diminishing returns applies. Overstaffing can be limited if in the past organizations developed people capable of being effective in more than one function of the business. It may be impractical to circulate a top-level scientist from function to function, but rotating people among design, development, manufacturing, marketing, and even finance has some significant benefits. Interdisciplinary or interfunctional expertise can be accomplished only by this type of movement. Staffing is undoubtedly the most important factor in determining the success of any project. The competence must be there to begin with, but staffing with overqualified people can be just as detrimental as staffing with the underqualified. There is always a limit, however, to the number of people who can be brought into a project at any one time and be used effectively. There are no set rules; everything depends on the business unit infrastructure. The major point to remember is that assigning people is only the first step toward meeting project objectives.

Graves' probabilistic approach must be modified. The assumptions must be questioned, and the distinction between research and development must be made. Although there are generally alternative solutions, seldom would a group have the luxury of fully evaluating 10 different approaches as Graves suggests. At best, there may be two or three, and the chosen approach may not be based on the technical considerations alone. The impact on the potential investment required in other functions may be an overriding factor. As an example, assume that two different technologies provide an appropriate approach but one requires a major investment in new production facilities. Which one would be selected? There is no simple answer. Many other system factors must be considered.

Viewing a project as a network of activities and then finding the critical path through that network tests a project manager's effectiveness. Finding the critical path depends on how critical path methods are interpreted. A research critical path will be quite different from a building construction critical path. Planning is limited to the known. That does not prevent research from planning, but the planning must be in the realm of the known. Graves mentions that there may be a tendency to select the easiest activities. That also tests the effectiveness of the project manager and the organization's senior management. Such a response would indicate a total lack of management concern about the project; it would define an inept manager. Expertise is required to select the appropriate and not the easiest activities. Selecting the easiest activities without determining the consequences would indicate a lack of managerial astuteness: wishful thinking rather than decisions based on the facts at hand. Selecting the appropriate areas for time compression are matters of judgment that involve knowledge of the situation and integrity in the decision-making process.

Graves provides information from the writings of others as to the percentage increase in cost for a 1 percent reduction in the project duration.

Hardware projects	1.75
Software, first source	0.88
Software, second source	2.00

Those figures indicate that a 1 percent reduction in duration can increase costs of the order of 1 to 2 percent; the percentages apply to a combination of research and development. If they are compared to those reported by Vesey [7], when real influence of product design wields more than a 50 percent influence downstream, the added costs to the project viewed from the systems and concept-to-commercialization perspectives may be minimal. At this stage of the project, the costs are essentially people costs.

Boston Consulting Group vice president Philippe Amouyal, in "Product Development as a Process" [9], in reference to the electronics industry, states that sound product development can result in resource savings of 30 to 35 percent. In the same article, Michael E. McGrath, a consultant, maintains that improving a company's development process can yield benefits of 40 to 60 percent. He cites Codex, a division of Motorola, in which product development cycle time was reduced from 140 weeks to 74.

The results from R&D can be extended to manufacturing, marketing, and the administrative functions. The case of the Portables Division of Tektronix (Chap. 10) is a good example. Not all the activities of manufacturing, marketing, and administration are routine. Each has research and developmental activities to support its operations. However, the same observations apply to the routine activities. The network of routine activities requires selecting the appropriate as contrasted to the easiest activities for time compression.

Assuming that time and cost projections are realistic and are based on practical expectations plus a stretch factor, project cost will most likely accelerate in the initial stages because of the need for added resources and the inability of an organization to change the working habits and relations that have become internalized. Management's ability to accelerate the change in attitude provides the greatest challenge in optimizing cycle time in all activities. The process used to manage the inputs to achieve certain specific outputs will determine the size of the area *EHJF* in Fig. 11.3.

Doing Things Fast—Pushing the Process

Schmenner, in "The Merit of Making Things Fast" [10], concluded that "focusing on throughput time forces managers to reduce inventories, setup time, and lot sizes: In addition it encourages improved quality, revamped factory layout, stabilized production schedules, and minimized engineering changes." Schmenner's research also shows that of all the possible techniques for improving productivity, only those related to just in time (JIT) are statistically and demonstrably effective. His research on productivity has linked throughput time reduction and other linchpins of JIT manufacturing to gains in factory productivity. Reducing the throughput time by 50 percent is worth an additional 2 to 3 percent in the rate of productivity gain.

Making things fast does not mean speeding up the machines, converting to a totally automated plant, or hassling the production workers for greater productivity. It means designing and organizing the manufacturing operations so that the raw material input moves through the system in a continuous manner without interruption rather than sit and wait in some aisle or temporary storage area for the next operation. In other words, it means eliminate the work-in-process inventory. The continuous flow of materials, coupled with well-maintained equipment that is free of intermittent breakdowns and reduced setup times, can add value for much longer periods. Schmenner is talking about what I like to refer to as pushing the process rather than pushing the people, keeping in mind that process and product are two sides of the same coin: They are a unit.

Schmenner's conclusion that "focusing on throughput time forces managers to reduce inventories, setup time, and lot sizes: In addition it encourages improved quality, revamped factory layout, stabilized production schedules, and minimized engineering changes" can be paraphrased to provide a generic principle that applies to every function within the organization. Focusing on throughput time forces all levels of management, beginning with the CEO, to reduce inventories of people at all levels (managing the ten resources according to the three E's), to eliminate non-value-adding activity, to eliminate the inventory of nonessential reports and so on, to reduce the total time that is required to complete all activities, and to integrate the business functions into a holistically operational unit. In addition, it encourages improved quality throughout the business unit, new flexible operational structures as required to meet divergent needs, stabilized operational schedules (people doing what they said they would do) in all functions, and continually decreasing the number of changes once the course of action has been established.

At first sight that paraphrasing from manufacturing operations to all of the business functions may appear to be an impossible objective. But it must be the objective if any significant change is going to occur in optimizing the three elements of cycle time: total time, timing, and cycle duration. It has already been noted that factory labor represents a smaller and smaller percent of total product cost. It is the costs associated with nonfactory employees that have grown rapidly and must be faced realistically. Vesey's comments in regard to the downstream costs that originate in design engineering [7] apply to all functions. Staff functions must begin to recognize the costs that they generate downstream of their activities. That applies to the CEO, every executive and manager, and all of the individuals who are in a position to assign individuals to specific activities.

Challenging Current Practice

Estimating *making time* or *doing time* in any area of business has not yet reached the level of a science. That is particularly relevant in introducing something new, something that has not been previously introduced. There are rules of thumb, but they apply to specifically well-described and quasirepeatable activities. They are usually extrapolated and corrected for differences with the past. Much estimating knowledge lies in the intellectual resources of the organization. Simple activities, when there is a history that can be referenced, present no problems.

Consider, as an example, the following situation of doing something that has never been accomplished, a true new-to-the-market product. *The Star Tribune* of Minneapolis [11] reported on the spokeless bicycle, called the "Zero Bike," designed by Makota Makita and Hiroshi Tsuzaki of Tokyo. There are no hubs and no spokes, and the bicycle is powered by cranking magnetic pedals that rotate tires suspended between other magnets. Those three conditions describe the new product. Your CEO or some senior executive calls you into the office and discloses this new design concept, provides you with some pertinent information, and, after some discussion of the pros and cons, asks you to prepare a proposal for the program. The executive has decided that the company will pursue the effort; it is not a matter of discussion or choice. You are expected to determine the critical R&D issues, the related marketing issues, a preliminary time line, required resource schedule, and the usual information that would be included in such a proposal. You accept the condition that the first draft of the proposal be completed in 4 weeks. Although this is not exactly a confidential request, it has not been given any wide publicity by your requestor.

After pleading with and begging functional managers for the required time from some specialists in the technology and marketing functions, you manage to accumulate enough information to begin developing a proposal. Four weeks later you are back in the executive's office and have gone through the ritual of making the formal presentation. You raised the issues related to the introduction of new technologies and the lack of adequate competence in the organization. You raised the issues presented by marketing. You outlined the critical issues related to costs, time, timing, and cycle duration. In addition, you presented your perceptions as well as the major issues that could negatively affect the project. The executive responds to your concerns and the critical issues that you have outlined. The estimated cost presents no problem, although past experience has taught the executive to mentally double or triple the initial estimated cost. This is not the first time the executive has looked at initial new-product costs. Experience has been a great teacher.

The executive has only one major concern: timing and cycle duration. Your estimated cycle time from concept to initial commercialization is 5 years. A whimsical look from the executive tells you that 5 years is too long. The company must be in the marketplace in less than 3 years. Cutting 2 years off the plan looks like an impossibility, but after some discussion you agree to review the program with all the supporting groups and provide a new estimate within 2 weeks. Mentally, you conclude that the executive must be reading the latest on time compression. What is your modus operandi?

Back to the drawing board, so to speak, more begging and pleading. Meetings with research, development, manufacturing, and marketing do not yield any significant breakthroughs. Some concessions that provide encouragement are made, but certainly not a reduction of 2 years. You begin reviewing the opportunities for reducing cycle time by activities that will and will not increase cost: taking the systems approach to the project, managing the interfaces, leveraging all the resources, focusing on value-adding activities, reviewing the limitations of the management system, eliminating the barriers, using multifunctional and multidisciplinary teams, simultaneous engineering, and the many other factors discussed thus far relative to optimizing total time, timing, and cycle time. After much deliberation, you conclude that 1 year could be eliminated if priorities could be changed in certain functional groups, if the essential elements of managing cycle time could be employed, and the concessions that you extracted from the functional groups were realistic.

Those actions only reduce the total time by 1 year. An additional year must be eliminated from the schedule. One avenue remains open: accelerating the project by adding resources. Adding staff raises additional issues. Can the additional people be integrated into the program effectively and in a timely manner? Adding temporary staff from contract houses, assigning people from other departments, and hiring new people are possible alternatives to reduce the cycle time, but can they be used effectively on this program? It is important to bear in mind that bringing new people into a project, regardless of the source, requires time to integrate them into the project and more time before they make a contribution.

You know that the costs will increase, and at the same time you are concerned about making decisions on critical issues without all the available information. After some thought, you discount the lack of information as a problem, because that is always the situation. If you waited for all the information, you could not accomplish anything, because more information would always be required. You conclude that "zero risk" is not a viable approach, and you set out to see what changes in your plan may be required and what the additional costs might be. What financial penalties will be involved to reduce the concept to commercialization cycle time to 3 years if the resources are available at the right time? The up-front costs related to R&D are relatively insignificant in the total product introduction process. The up-front costs involve mostly people costs and prototype work. However, manufacturing would require new equipment, to the cost of which a premium for timely delivery must be added. The new equipment requires some research, since it cannot be purchased off the shelf. It involves not only new manufacturing technologies related to this project but technologies that have not previously been employed successfully.

After considering different alternatives, you select two that are considered to be achievable. The timing would be difficult but achievable if all the resources are available. You are apprehensive and raise additional questions, because the literature on reducing cycle time suggests contradictory evidence. Some say it doesn't cost more. Others state that each 1 percent reduction in cycle time adds at least 4 to 5 percent to the cost. Others tell you what great benefits you will receive by introducing a product earlier than projected. If the objective is to reduce the time by 1 year solely by acceleration, that could mean an increase of 100 percent in the R&D cost. You ignore all the statistics and the published research and make one more trip to the accounting department to reconsider the time-cost trade-offs from a total business perspective. Your colleagues in the accounting department are not particularly glad to see you, but they will help evaluate the financial aspects of your proposal, the business cost-benefit analysis, the return on investment, the return on capital employed, the cash generated for the business, and so forth. In addition, they examine the worst-case scenarios. You are now ready to present your proposal. Several additional iterations may be required before a workable proposal is developed—that is part of the up-front work. In the meantime, the executive cannot understand why it is taking so long to develop a proposal that is acceptable to the technology functions, marketing, and the accounting department. Unfortunately, that is the way most projects originate. Such an approach has severe limitations. First, the complexity of the problem is completely underestimated by the executive. The individual assigned to the activity is naive and lacks the fortitude to confront the executive with the impossibilities of achieving a realistic answer. The functional managers have not received any clearcut signal about its importance. The critical mass of competence has not been assembled. The functional disciplines operate in isolation.

The major purpose of this scenario is to illustrate that the decisions relative to managing cycle time of a new-to-the-market product involving new technologies and new markets cannot be undertaken as some type of add-on activity. It is not just a matter of adding more cost to shorten the cycle. Each aspect of a major project must be evaluated on its merits—its contributions to the total business. Cycle time can be reduced in a combination of ways that include reevaluation of concepts and the resources to implement them, integration of the functions and acceptance of accountability in relation to the total system, and attention to the interfaces, the value-adding activities, and the other barriers that not only accumulate hours but extend the duration.

Such scenarios consume resources and, perhaps more important, extend cycle time because of the negative impact on employee morale. How many more times are we going to go through the same thing? Why can't top managers make up their minds? Top managers really do not know what is involved in this particular scenario. They lack an understanding of what is involved in introducing new technologies, creating a new product, and entering a new market (Chap. 5.).

Consider the same request for a proposal regarding the spokeless bicycle. Managing the three elements of cycle time is *activities-based management*. The three-dimensional model introduced in Chap. 2 related the infrastructure and the resources to the business activities. Whether an activity is described as that multimillion dollar project related to a new-to-the-market product or the thousands of activities within that project is immaterial. The resources and the infrastructure must be known and must be in place. Without those resources and without a supportive infrastructure, which determines the effective use of the resources, cycle times are extended. The infrastructure-resources-activities model provides a structured approach for analyzing any business activity considered for investment. If an organization has a good grasp of its "as is" condition, the limitations of the three components of the model can be developed easily and quickly. Only a dose of "integrity pills" is needed to bring all the participants into the real world. The added cost depends on how many of the elements of the model are in place. If the guiding principle of the organization is to disregard the need for doing the up-front work diligently, cycle time will be extended. If an attempt is made to accelerate cycle time under such conditions, costs could escalate rapidly because the lack of the up-front work will generate additional costs that are due to changes and additional rework throughout the whole process.

Even though teams are currently a popular topic in the business and academic literature, they have been around in one form or another for centuries. There is no doubt that complex systems that involve many disciplines require more integration than a system that can be described by only one discipline. How an organization chooses to describe the team and when it organizes the team make a considerable difference in performance. Personal experience has demonstrated that projects of long duration (3 years or more) should be preceded by some form of exploratory committee or task force that (1) identifies the issues, (2) identifies the scope of the project, (3) performs the necessary analysis, (4) develops the complete project proposal and spells out the specific needs, alternatives, pros and cons, resources, timing, and so on, (5) proposes the core team, (6) gains approval of the project, and (7) takes responsibility for performance. In the example of the spokeless bicycle, the first four activities were given to one individual without any emphasis from senior management as to its importance.

Identifying the issues (doing the up-front work by an exploratory team), which translates into clearly delineating and describing the many factors involved in a major project, cannot be left to chance. The level to which the up-front work has been performed determines the number of subsequent changes. The purpose of the original exploratory group is not limited to developing a project statement; it includes fully understanding the requirements and thinking before doing. In the process of the group's deliberations, some experimental work may be necessary to verify its conclusions. Discussions about the marketplace would also be involved. All the elements of the system that can be identified must become part of the exploratory process. The major issues that could limit successful implementation must be identified before a formal project is approved. Depending on the size and scope of such an endeavor, it is beneficial to establish the exploratory effort as a project. Since the project involves cooperation from a multiplicity of functional groups, those groups must provide their expertise in preparing the proposal. Before the project could be approved, intelligent management would want to know the extent of the resources to be allocated to the new effort—not just the financial resources but, more important, the people resources. Unless the activity is designated as a senior management level effort, it will be difficult to gain the support of functional managers that may consider themselves already overloaded with work.

Assuming that the project scope and all the other details and assumptions are based on reality, selection of the core team determines project success. The individual who leads the exploratory group should be the one capable of becoming the project manager, and some of those on the original exploratory team should be assigned to the core team. Core teams must be autonomous, but with appropriate checks and balances. An active and involved senior management level steering committee also provides direction once the project has been approved. One of the most difficult tasks of the project manager is to keep the team directed to the approved objectives.

The knowledge base for any activity must be uncovered and understood before an organization can hope to affect its cycle time on a specific project. In that sense, optimizing project cycle time or time to market is determined entirely by the management process. The question whether cycle time must be shortened becomes irrelevant. It is the process by which an organization manages its activities to optimize the total time, the timing, and the duration of the cycle. Success depends on so managing the activities as to achieve the expected results from assigning resources to the same activities. Managing for results without knowing how the results are being accomplished naively accepts as fact that results alone determine future performance.

Recent emphasis on process has directed many organizations down the wrong path. A multiplicity of programs related to quality, empowerment, continuous improvement, benchmarking, and so on, if implemented solely through an emphasis on process, are bound to have little impact on organizational performance. The results may include an excellent quality program, a program for empowering the organization, another program for continuous improvement, and a group that understands how to develop benchmarks against which to measure performance. Whether quality will be improved, people empowered, continuous improvement implemented, and benchmarking used for business advantage is questionable. Process must be tied to activities directed toward results.

Schaffer and Thomson, in "Successful Change Programs Begin with Results" [12], state that: "Most corporate change programs mistake means for ends, process for outcome. The solution: focus on results, not activities." To partially justify their conclusion, they cite a 1991 study by the American Electronics Association of 300 electronic companies: 73 percent reported having a total quality program but 63 percent of them failed to improve quality by 10 percent. There is no doubt that focusing solely on activities will not provide a significant benefit. But neither will swinging the pendulum to the other extreme—results. Results are achieved through specific well-described and well-integrated activities. Activities are accomplished through processes, an understanding of the processes, adequate control of the processes (balance open loop and closed loop), and focusing the activities on results. Activities, processes, and results cannot be separated and treated as independent variables. The assumption cannot be made that activities will yield results. Educating every person in the organization on the latest single-issue management panacea only provides education. That education by itself yields nothing in improved performance unless there are some definitive expectations established as a result of that education. Likewise, focusing on results through single-issue programs without well-described activities that integrate the functional entities will not yield acceptable results. It is this type of single-issue pursuit that prevents organizations from improving performance. Activities, processes, and results are one. As a unit, they determine the throughput from allocating business resources to fulfill the purposes, objectives, and strategies of the organization.

Justifying the Selected Alternative

The costs associated with accelerating cycle time must be weighed against the potential benefits. It would appear that such a statement is a given, a truism, yet organizations fail to apply the fundamentals associated with evaluation of alternative investments. They place a significant emphasis on investment justification. Gaynor [13] made a distinction between project justification and evaluation of alternatives followed by selection of the most appropriate alternative. The time spent on justifying a predetermined approach generally exceeds the time required to evaluate the alternatives. Trying to make the figures come out right can consume a great deal of time without providing any insight or understanding. Usually more than one approach to an investment can be taken. Some can be discarded immediately for any number of reasons. The difficulty arises when two or three viable alternatives may provide very different levels of investment with correspondingly different benefits and the decision maker insists on some preconceived approach that may not be the most viable alternative.

Figure 11.4 suggests six key issues that must be resolved before justification of any alternative. Selecting the appropriate alternative that meets the preestablished criteria would appear to be a routine process: adding up the yes and no columns. But the answer is not that simple. Most likely, no alternative will meet all the criteria. Managing the trade-offs requires a systematic approach. Weighting the issues related to technology complexity, market penetration, manufacturing cost and process complexity, and other considerations does not lend itself to complete quantification and a neatly designed decision process. The selection process must take into account the fact that at this stage, although numbers are presented as fact, they are merely numerization of

1. Understanding the purposes and objectives of the business and the strategies at all levels and within all functional groups that will allow the purposes and objectives to be achieved.

2. Knowing the limits of and the availability of the ten resources within as well as outside the immediate business unit and understanding the limitations of the supporting infrastructure.

3. Analyzing and synthesizing the technology requirements related to the specific product. The *paralysis by analysis* syndrome dominates much of business decision making. Synthesis is forgotten. All the analyses from the functional groups must be synthesized into some kind of a meaningful whole; the pieces must be reconciled.

4. Designing the change scheme with the alternatives. The alternatives must be clearly delineated. All the essential functional groups must participate in development of the alternatives. The preliminary investigation for an investment cannot be conducted in isolation of the supporting functions.

5. Evaluating the alternatives requires establishing the criteria before the evaluation. Developing criteria is not a simple process. The emphasis should be placed on the criteria that are essential for consideration. If the specific criteria are not met, the alternative is not acceptable.

Figure 11.4 Key issues that must be resolved prior to justifying an acceptable alternative from a series of opportunities.

qualitative information. In the final analysis, the decision will be a matter of judgment—most likely one individual's judgment.

The accounting principles related to financial justification of projects have been essentially cast in bronze. Creative accounting may be a new field in the 1990s, but creativity in accounting practices has not as yet reached anywhere near its optimal potential. As an example, arguing with accountants about the misapplication of discounted cash flow quickly leads to a dead end; supposedly, there is no other way to justify an investment. After considering the shortsightedness of applying discounted cash flow to all investments (research is an investment), a colleague sent me an article by Thomson: "Dangers in Discounting" [14]. Thomson states that there is an equivalence between discounting dollars and discounting time. Since many organizations arbitrarily use a figure of 15 to 20 percent in discounted cash flow calculations, projects that return the investment in the shortest period of time receive priority. However, a lower discount rate may provide a much larger cash flow over the total life of the investment. The present value of $1 per year for 3 years (discounting dollars) is the same as $1 per per year for 1.95 years (discounting time). Comparison of the two approaches, discounting dollars and discounting time, shows a dramatic difference in time horizons between discount rates of 20 and 5 percent. At the 20 percent rate, the first 20 years occupy 98.2 percent of the time horizon. At

5 percent discount rate, the first 20 years occupy 63.2 percent of the total time horizon.

Investing in technologies that result in new products challenges business managers to manage risk and uncertainty. Consider the situation of Supercomputer Systems, Inc. (SSI), which is challenging Cray Research in the supercomputer market [15]. In essence, designing and building a supercomputer is a new-product project with many subprojects. SSI was founded in 1987 by Steve Chen, a former designer at Cray Research. He supposedly left Cray over basic disagreements about the next generation of supercomputers. The company spokesman insists that SSI is on course and will introduce its supercomputer in the early 1990s. Some computer analysts are concerned that the production schedules may be slipping from 1993 to 1994. One industry analyst suggests that SSI is falling behind its timetable. He claims that potential customers indicate the shipping schedule has slipped a year, putting it in 1994.

Developing a new supercomputer is certainly no minor activity. It requires vast amounts of creativity, ingenuity, innovation, resources, and application of management skills. The SSI case illustrates the impact of cycle time. SSI started operations in 1987, and 7 years or more will be the time from concept to initial commercialization. A market that was once dominated by Cray Research now includes the Japanese computer makers NEC, Fujitsu, and Hitachi. The major unanswered question is what the supercomputer market will look like when SSI ships its first unit. Whether or not Steve Chen considered the impact of cycle time on product entry is unknown. What is certain, though, is that he had only a concept; he did not have a design. In the interim period, applicable technologies have changed. Success in this case will be determined by the concept to commercialization cycle time. Competence in technology, although essential, is insufficient.

Summary

Estimating and measuring the cost of accelerating cycle time appears to be a routine matter. The calculations may be routine, but the information on which the calculations are based is often speculative and the events are unpredictable. That is the nature of business in a free-market economy. Assuming that management and the decision makers understand most of the aspects and idiosyncrasies of their business, the activities in determining whether to accelerate the cycle time are routine. Risk may be involved, but evaluation of that risk is part of the process. The original risk of investing in the project should have been determined when it was first considered. The only question that remains to be answered is what additional risk is associated with accelerating the cycle time. Assuming that estimates of time, cost, and performance were realistic, reducing the time will add cost. In the short term, management must satisfy itself that the additional cost can be recovered and in the process provide some additional benefits.

Determining the correct cycle time must go beyond making commitments and then attempting to determine how to accelerate the process. The process must begin at the time the concept is being formulated, recognizing that optimizing cycle time also includes time and timing. Good management practice would indicate that all projects, regardless of origin, purpose, or content, take into account the elements of time, timing, and cycle time. Managing the three elements of cycle time by the three E's is not a matter of choice.

References

1. S. G. Shina, "Concurrent Engineering—A Special Report," *IEEE Spectrum,* July 1991, pp. 22–41.
2. P. B. Crosby, *Quality Is Free,* McGraw-Hill, New York, 1979.
3. Christoph-Friedrich von Braun, "The Acceleration Trap," *Sloan Manage. Rev.,* Fall 1990, pp. 49–58.
4. F. J. Contractor and V. K. Narayanan, "Technology Development in the Multinational Firm: A Framework for Planning and Strategy," *R&D Management,* vol. 20, no. 4, 1990, pp. 305–322.
5. E. Mansfield, "The Speed and Cost of Industrial Innovation Japan and the United States: External vs. Internal Technology," *Management Science,* 1988, pp. 1157–1168.
6. E. Eckard, "Alternative Vertical Structures: The Case of the Japanese Auto Industry," *Business Economics,* October 1984, pp. 57–61.
7. J. T. Vesey, "The New Competitors: They Think in Terms of Speed-to-Market," *Academy of Management Executive,* vol. 5, no. 2, 1991, pp. 23–33.
8. S. B. Graves, "Why Costs Increase When Projects Accelerate," *Research/Technology Manage.,* vol. 32, no. 2, March-April 1991, pp. 16–18.
9. R. Whiting, "Product Development as a Process," *Electronic Business,* June 17, 1991, pp. 30–36.
10. R. W. Schmenner, "The Merit of Making Things Fast," *Sloan Manage. Rev.,* vol. 3, no. 1, Fall 1988, pp. 11–17.
11. "Hiring Plans Cast Doubt on SSI's Growth Projections," *Star Tribune* of Minneapolis, Aug. 16, 1991, p. 10 A.
12. R. H. Schaffer, "Successful Change Programs Begin with Results," *Harvard Bus. Rev.,* January-February 1992, pp. 80–89.
13. G. H. Gaynor, *Achieving the Competitive Edge Through Integrated Technology Management,* McGraw-Hill, New York, 1991, pp. 165–185.
14. C. T. Thomson, "Dangers in Discounting," *Management Accounting,* January 1984, pp. 37–39.
15. *Star Tribune* of Minneapolis, Aug. 20, 1991, pp. 1D–2D.

Case Histories and Corporate Applications

Learning or thinking about the fundamentals adds value up to a point; applying those fundamentals presents the real challenge. Knowledge of successful and unsuccessful applications also can ease the path toward implementation. Throughout this book, case histories have been presented to illustrate specific points and demonstrate what has been accomplished by some organizations. Chapter 12 expands on those cases and presents new histories that show the limitations and the potential as currently perceived by organizations. These case histories will clearly illustrate the points made in the preceding chapters. Reducing time to market involves not only integrating the technology and product-related functions but the whole organization.

Learning from Others

Case histories seldom reveal detail sufficient to make replication in a different environment possible. That is especially true of an attempt to validate the contribution of some specific set of activities. The total result can be measured, but isolating the specific activity or set of activities always leaves some questions unanswered. Personal bias and preconceived notions of the relevance of certain parameters also detract from systematic analysis. That applies to technologically related as well as management-related activities. As an example, manufacturing productivity is rooted in the ability of the prevailing infrastructure to allocate and leverage the available business resources to the activities that add value. Assuming that the three components meet the business requirements, organizations attempt to improve performance by promoting issue-directed programs. Some organizations place all such activities under some version of the quality umbrella. Others may focus on advanced manufacturing technologies. Cost can be a focused program. Cycle time or speed to market is another. All of those programs are closely linked. Attempting to allocate productivity or improved performance to one specific program may be a waste of time and effort. A reduction in cost of quality is most likely associated with new production processes, changing market conditions, improved product design, and elimination of many of the management barriers to cycle time discussed in Chap. 9.

Figure 12.1 illustrates the relations of organizational effectiveness and time delays. As shown, the business system functions somewhere in the grid. The curve AB shows that effectiveness decreases as time delays increase. The objective must be to drive the time delays generated in the business system as close to zero as possible. The time delays occur in all segments of the organization. The reasons for the delays and the approaches taken by different organizations to resolve the delays are discussed in the following sections: Lead Time in the Auto Industry, Contrast between Canon and Apple, Corning—Competing in

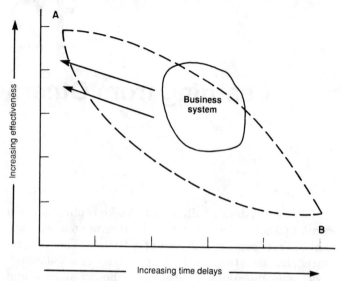

Figure 12.1 Relations of business system effectiveness and the sum of the time delays.

Video Glass, General Electric, Hewlett-Packard, IBM, Motorola, NEC Corp., Northern Telecom, Target, Thornton Equipment Company, Toshiba, Velcro, and Use of Special Tools.

Lead Time in the Auto Industry

Comparisons between U.S., European, and Japanese automakers continue to be refined as researchers begin comparing modes of operation and attempting to rationalize the differences. Clark and Fujimoto, in "Lead Time in Automobile Product Development Explaining the Japanese Advantage" [1], focus attention on development lead time—the time from concept to market introduction. They frame the product development process in the auto industry around four activities: concept study and product planning as the *planning activity* and product engineering and process engineering as *engineering activity*. They have identified four factors that affect lead time:

- *Work load.* The product content—affects planning and engineering
- *Capacity.* Ability to solve problems quickly, integrate multiple activities, and pursue necessary activities in parallel
- *Planning.* Activities that lay the foundations for the engineering effort

TABLE 12.1. Selected Findings by Clark and Fujimoto
Regarding Differences in Lead Time in the Auto Industry

	Japan	U.S.	Europe
Overall lead time, months	43.3	60.3	66.5
Planning lead time, months	13.6	19.3	20.6
Engineering lead time, months	32.2	39.7	45.9
Die development lead time, months	13.8	25.0	28.0
Prototype lead time, months	6.2	12.4	10.9
Ratio of carryover parts, percent	18	45	29

- *Engineering.* Simultaneous problem solving among interdependent activities such as linking design with tooling, finishing, and any activities that affect the component or assembly

Each of the four activities involves its own lead time and integration of different levels of managers and professionals. The lead times are affected by product content, project scope, and organizational competence. Table 12.1 presents some selected findings by Clark and Fujimoto. Japan's total lead time is 17 and 22 months shorter than in the United States and Europe, respectively. Japan's planning lead time is 6 months shorter and its engineering lead time is 7 months shorter than in the United States. Lead time is affected by content, scope, and organizational competence. The Japanese introduce fewer new concepts but more new body types. *Fortune* reports that Toyota captured a 43 percent share of the Japanese market by introducing six all-new vehicles in 14 months [2]. Toyota needs only 13 worker-hours to assemble a car. It has a network of 228 suppliers that produce jigs and fixtures, do research and development for Toyota, and provide a wide array of components. That allows Toyota to leverage its resources. Toyota also uses a higher percentage of new parts. The authors found that project scope greatly impacts planning time but not engineering time. Organizational competence significantly affects engineering time but has little impact on planning time. Clark and Fujimoto conclude that although the raw data give the Japanese a 19-month advantage, factoring in the project content and scope gives the Japanese a real advantage of 12 months. That advantage comes from internal organizational competence, a strong network of suppliers, and an innovation strategy that focuses on many continuous incremental changes.

There is no research that clearly documents the performance benefits of short lead times—that is, its direct and independent contribution to bottom-line performance. If viewed from the perspective of managing an organization effectively and efficiently, and in the process including the three elements of cycle time (total time, timing, and cycle duration), reducing the lead times is bound to have a positive impact.

Contrast Between Canon and Apple

Cycle time and innovation are interconnected. Innovations that cannot be brought to market in a timely manner do not provide any added value. The exception may be the case when the innovation involves a new-to-the-market product and provides the impetus for a totally new business. Innovations in existing products for improved reliability or expansion of capability are dominated by the three elements of cycle time.

Nonaka and Kenney, in "Towards a New Theory of Innovation Management: A Case Study Comparing Canon, Inc. and Apple Computer, Inc." [3], discuss the differences in management approach to innovation at Canon and Apple. Canon introduced a plain-paper copier (PPC) in 1969 with technologies that did not violate any of Xerox's 600 patents. In 1982, Canon began reconceptualizing the PPC business and decided to investigate the opportunities for lightweight, compact copiers. It was stipulated that the technological problems would not be resolved by applying some brute force methods or just improving current designs. Canon approached the project with a task force appointed by the company president. The list included:

- Task force chief—director of the reprographic products development center (RPDC)
- Advisor—managing director of the RPDC
- Quality assessment and cost assessment groups
- Market task group
- Soft task group to examine color copying, and so on

Kei Saito, the deputy general manager of the Advanced Technology Development Department of RPDC summed up the philosophy that would guide the project:

> In any company good products are created when production engineering and design become fused for their development. I believe there were tangible and intangible cost reductions. By becoming one with Production Engineering, one can propose uniform parts design, or assembly in one direction, how something should be assembled and in what sequence, or that one should do this or that if possible, when attempting to automate production, for example. If we in [product design] are by ourselves, it is easy to prepare drawings, and do what we like without thinking that far ahead. So, our discussions with Production Engineering people and working to accommodate their various requests in our own ways resulted in both tangible and intangible cost reductions.

Fusion of all the related disciplines, needs, and wishes of different groups is not a simple process. Whether such a group is designated as a task force or team of some type is unimportant. Such actions lead to major synergistic and cumulative effects on performance. It is interesting to note the high levels of management involvement. This was obviously a major effort by Canon, but the top levels of the organization were involved. When a project receives such top-level participation, the necessary resources will be available. Communications involved the managing director of research and development, director of the corporate technical planning and operations center, and the director of the production engineering research center. Perhaps this is one of the conditions for exploiting synergy in managing cycle time. As is so common in many Japanese organizations, Canon emphasized doing the up-front work with a great deal of care.

It is interesting to note that Canon, on entering the PPC business, managed to work around 600 Xerox patents. The value of those 600 patents to Xerox must be questioned. There is a great deal of discussion of the value of patents in assessing the output of a research or development organization. This tends to reinforce my biases that numbers of patents are not much of a guide to performance. It is more important to emphasize the substance of the patent. It may be more important to be the owner of one patent on a new material of construction than 100 patents on mechanical or electronic configurations that can be circumvented with some minor changes.

The organization at Apple Computer, Inc. was more fluid. Here was a start-up company that began in a garage (1976) with an idea (Apple II) and in 5 or 6 years (1982) was generating sales of $750 million. The market did not accept the Apple III and the Lisa, but the company was cash-rich. Management continued in the hands of the founders, and there was little financial or operational discipline. Nonaka and Kenney describe Apple as in a state of confusion with no direction, and too many projects. Working on too many projects and without adequate resources is a common U.S. industrial practice.

An attempt to turn Apple around began in 1979. Since Steve Jobs was no longer on the Lisa project, he became attracted to the Macintosh group. By 1981 Jobs replaced the original project leader of the Macintosh project and began to build a project team—he became the product champion. Many features were added to the specifications of this smaller, faster, more powerful, and less expensive machine. Jobs recruited people from Xerox Parc and even used some of the 1970s technology. The project was not exactly structured with a clear or precise idea of its final configuration. A development schedule did not exist. Supposedly problems and solutions were crystallized simultaneously. Jobs set out the basic principles; the team of 25 people brought them to fruition. The team was physically separated from the rest of the organization, and total involvement and continuous communication and interaction were possible. Jobs was the product visionary, the champion, and the final decision maker. The success of the Macintosh in its many forms is history.

Canon and Macintosh are presented as two distinct ways of bringing a new product to market. Canon was well integrated throughout the organization: a unified top-management approach, agreed-upon objectives, target dates that were met; 600 Xerox patents that were circumvented, and availability of total resources of the company. At Apple, the effort was divorced from the major operations and isolated; management was initially ambivalent; Jobs had the vision; the objectives were not clearly defined in detail; limited resources were available. Since Jobs was the final arbiter, decisions were made on personal bias and his own idiosyncrasies. The Macintosh was a major advance in personal and user-friendly computers by Apple. Sales by Apple eventually reached $4 billion. However, the transformation at Canon and Apple were quite different. After the Mac project was completed, much of the talent was dispersed. Even Jobs left to start NeXT Computer. At Canon, that mass of talent remained. What was learned in the process of bringing the minicopier from concept to commercialization became a human and intellectual resource that could be leveraged in the future.

Nonaka and Kenney presented their analysis in relation to a new theory of innovation rather than examples of managing cycle time. However, the examples illustrate two totally different styles of bringing a new product to market. Canon focused on the up-front work, developed a corporate program, provided the essential resources, established objectives, and so on. Apple with its Mac project grew like Topsy, emerged from a somewhat chaotic environment, finally received management's support, and was very successful. No information about cycle time is available, nor would it be appropriate to compare the time to market. However, Apple went through a series of false starts, organizational changes, and, because of its autocratic leader, probably ignored many good ideas. Moritz, in *The Little Kingdom* [4], states that: "False starts, diversions, mistakes, experiments, rebellion, and competition formed the stuff of the machine." There is no doubt that, under such conditions, total time increases and the project duration is extended. If Jobs had been required to design a system that was outside a competitor's 600 patents, it is most likely that the Mac never would have reached the marketplace.

Corning: Competing in Video Glass

Teresko, in "Corning's Rebirth" [5], discusses the outcome of critical decisions made by the company in 1985. Does it make a major investment in color television glass manufacturing and compete with the Japanese or does it exit? The decision: Stay in. This Corning case history has all the elements related to cycle time management and its impact on business performance. It also has all the elements related to innovation, introduction of advanced manufacturing technologies, demand for improved quality, and even a joint venture with a Japanese company. Corning grew its television glass business in an environment in which quality was not a major issue—the market needed vast numbers of TV sets. When most U.S. TV manufacturers exited the business, that created problems for Corning. The Japanese TV industry favored its own suppliers. Corning abdicated the market in 1975 when it closed its last black-and-white video glass plant in the United States. By 1985, it operated one color video glass plant. Corning management had to make a decision: Quit or fight. It decided to pursue the market although significant risk was attached to the decision. It was a $3 billion market. Corning adopted a three-point strategy:

- Marketing strategy to penetrate the marketplace

- Manufacturing strategy to be a world class competitor—emphasis on process

- Work place strategy to change the embedded organizational culture

In 1988, it formed a marketing approach with Asahi Glass Co. Ltd., Tokyo. Corning has a major interest in Corning Asahi Video Products Co. (CAV). The quality problems are being solved in production, the American way. The Japanese approach is to rely on employee ability to micromanage the process; CAV collects data from 6500 different points in the production process. Fulton, the manager of manufacturing and engineering planning, is convinced that process knowledge is the single most important success-related factor. The current process still takes 2000 h of process time to achieve yields of 60 percent. The objective is to reduce the time to 200 h. The change in culture has been the most difficult, and it took more time than anticipated. Middle management had as much difficulty changing as any other group of employees. Now it is difficult to distinguish engineers from operators. Education and training involves about 5 percent of the time on the job.

The reorientation of the business has also brought advantages to the design operations. Making the molds for the specific sizes and shapes of video glass can be a long and detailed process. With specialized computer-assisted design software, the mold design time has been reduced from 10 weeks to 1 week. Wendell Weeks, CAV's business manager, describes the future as follows; "In five years, if I had my way we'd have no sales force or account managers—just direct plant to plant interaction from us to our customers." By 1989, gross margins were 3 times those in 1985, manufacturing cost was down 22 percent, quality was up by 42 percent, inventory turns had tripled, and product development time was reduced by 50 percent.

General Electric

Throughout this book I have stressed that single-issue management neglects the fundamentals of managing and that cycle time, although it must be brought to the front burner, is a result of effective management practice rather than a new program. Tichy and Charan, in "Speed, Simplicity, Self-Confidence: An Interview with Jack Welch" [6], discuss some of the changes being pursued by General Electric's CEO. Welch's comments provide all levels of management with some reflections on improving performance; achieving the required cycle times is one of those elements. His comments illustrate reducing time at all levels of the organization.

Welch prefers to think in terms of business leaders rather than managers. He differentiates the leader from the manager as one who creates a vision, articulates it, passionately pursues it, and drives it to completion. Leaders are flexible, informal, straight with people, accessible, and candid in their communication. They communicate in the real sense rather than by announcements and pronouncements. They face reality. They see the situation as it is rather than as they would wish it to be.

What has Welch done that affects the elements of cycle time? First, the organization has been simplified. Although many CEOs consider having 6 or 7 people as a maximum of direct reports, Welch considers that 10 to 15 is preferable. It reduces the questions, eliminates the second guessing, and limits the time available to managers for checking and nitpicking. General Electric consists of 14 separate businesses. A major investment decision can now be made in days. In one situation that involved the swapping of GE's consumer electronics business for another organization's medical equipment business, it took 30 minutes to determine if the deal made sense and a 2-h joint discussion to work out some basic terms. A letter of intent was signed in 5 days. The total length of time, which included working out the usual legal details, took 5 months.

As part of an effort to simplify the organization (communication) Welch formed a corporate executive council (CEC) that meets every quarter for 2 days. It includes the leaders of the 14 businesses and top-level staff people. These are not formal strategic reviews; they are opportunities to share ideas and information about the failures as well as the successes. At a 1986 officers' meeting, the leaders of the 14 businesses were asked to present a one-page response to five questions related to global market dynamics in the next several years:

- What are the global market dynamics today, and where are they going?

- What actions have your competitors taken in the past to upset those global dynamics?

- What have you done to affect those dynamics?

- What are the most dangerous things your competitor could do in the next 3 years to upset those dynamics?

- What are the most effective things you could do to bring your desired impact on those dynamics?

The outcome of such a discussion is not a ring binder consisting of extrapolated future business projections, but as Welch states, "a playbook with five charts for each of the fourteen businesses. We know exactly what makes sense; we don't need a big staff to do endless analysis." Through such an approach the business leaders can devote their energy to creating and building new businesses and the executive offices can function as facilitators in the process and transfer the best practices across the different businesses.

General Electric is known for its emphasis on education at all levels. Part of that management education process involves Work-Out. Work-Out focuses attention on how things get done. It has a practical and an intellectual goal. The practical involves getting rid of the thousands of bad habits accumulated since the organization of General Electric, work habits that include the policies and procedures and all the excess baggage that no longer provide a qualitative or quantitative benefit. The intellectual part of Work-Out involves 8 to 10 meetings per year by the business leaders with their people. The purpose is to expose the leaders to what is really going on and how the participants view the many issues that affect performance. That is part of Welch's program on candor. Its purpose is to redefine the relations between the boss and the subordinates. The thrust of the sessions is to get people to challenge their bosses about why so much wasteful effort is required.

Tichy and Charan illustrate the Work-Out approach with a case study at GE Medical Systems (GEMS). This group involves diagnostic imaging and x-ray mammography; it includes about 15,000 employees. The GEMS Work-Out session was designed to identify the sources of frustration, uncover the inefficiency created by the bureaucracy, eliminate unnecessary and unproductive effort, and find new ways to reward managers. Work-Out is an effort to unravel, evaluate, and consider the complex web of personal relationships, cross-functional interactions, and formal work procedures through which people accomplish objectives—through which they make a contribution to the business. It ends with development of written agreements by the participants to change the way people work. It is an action plan for a transition to new behaviors.

The GEMS Work-Out involved about 50 people including John Trani, a senior vice president and a group executive. Trani selected people from technology, marketing, and so on that he thought were willing to take risks, challenge the status quo, and make a contribution. The Work-Out session was preceded by in-depth interviews with managers at all levels and a visit by Jack Welch to the GEMS headquarters for a half-day session with the participants. The in-depth interviews brought out some of the issues related to procedures, systems of measuring performance, cross-functional conflicts, the groups total environment (attitudes, feedback, and so on), lack of team commitment, and the issues related to careers. Since GE was in the process of downsizing, one respondent said:

> The goal of downsizing and delayering is correct. The execution stinks. The concept is to drop a lot of less important work. This just didn't happen. We still have to know all the details, still have to follow all the old policies and systems.

Welch's participation in the presession yielded some very common assertions. Participants stated that top management:

- Focuses on too many preconceived beliefs—beliefs, not facts
- Spends more time pontificating than listening
- Attempts to second-guess lower-management levels

The GEMS Work-Out session yielded almost 100 written agreements between individuals, between functional teams, between department heads and their staff, and so on. Such sessions obviously provide a benefit if the expectations are within the resources and a follow-up system is developed to track performance. This tracking system is not another lengthy report; it is only a matter of whether the agreement was or was not honored and what the benefits were. In other words, what has changed because of the agreements? Implementing change takes time, but change can be accelerated with such programs as Work-Out. Although the need for change can originate at the top, it will be implemented from the bottom up. The process can be accelerated by getting the people who can create the change involved. There will be difficult decisions along the way, as GE found. Some people may just have to be removed from the system.

The GE case offers other examples related to the elements of cycle time. Sherman, in "The Mind of Jack Welch" [7], notes that at one time the company's finance department revealed that some 1000 people worked nights and weekends at the end of every quarter to ensure that GE was the first company its size to publicly report its earnings. Welch eliminated the emphasis on being first to report: It did not add any value. The example illustrates how significant amounts of effort can be devoted to totally useless activities. Every organization is faced with similar time-wasting activities that affect the three elements of cycle time and only add overhead cost to the cost-benefit equation. In Chap. 6 the changes implemented at GE's Electrical Distribution and Control Business provide an additional example of an organization reaping considerable benefits in cycle times by managing the business effectively and efficiently.

Hewlett-Packard

Arguing for renewal in manufacturing continues to be a topic of discussion in industry. The attitude and belief that business performance could be improved if manufacturing got its house in order prevail to a great extent. That translates into a better-quality product at a lower cost and in a shorter period of time. In recent years, many new manufacturing soft tools made their appearance. Mentioned in this book are CAD, CAM, CAE, CIM, MRP, MRPII, JIT, TQC, SPS—just part of a seemingly never-ending list of acronyms that will transform manufacturing into a world-class operation. Hewlett-Packard (HP) has been pursuing superiority in manufacturing for over a decade. Quality, cost, product and process integration, business systems thinking, and manufacturing's involvement in R&D have been well documented. The objective was that the integration of activities would become the rule rather than the exception. Edmondson and Wheelwright, in "Outstanding Manufacturing in the Coming Decade" [8], describe the current program at HP to speed up the pace of change.

HP's manufacturing is guided by a manufacturing council consisting of a corporate vice president and the senior managers of the business groups. The council identified nine dimensions that provide the foundation for accelerating change within the manufacturing operations: people at all levels, manufacturing strategy and plans, goals and measurements, teamwork, organization, information systems, physical processes, machinery plant and equipment, and flexibility. HP's council also proposes that competitiveness in manufacturing depends upon the multiplicative and not merely the additive effect of the related activities. However, the nine areas that govern manufacturing's effort toward superiority apply to the rest of the organization also. HP approaches its activities not only with those nine principles but with an understanding of the contributions required from manufacturing and its supporting functions. A decision was made to transform its divisional manufacturing operations into group level to take advantage of economies of scale. The fact that manufacturing became more visible through that process allowed senior-level manufacturing management to play a significantly greater role in initiating actions related to changing corporate time-consuming systems.

Generally, performance improves when an organization learns how to implement a continuous flow of minor improvements. Installing a new, automatic line may be a major endeavor, but implementation and the commissioning are only the beginning. The real benefits come about from a concerted effort to make minor improvements that collectively allow approaching the theoretical limit. HP's Computer Systems Group realigned its manufacturing operations with the idea of "cheap to change." That involved so designing the facility that the layout could be changed over a weekend. It was coupled with developing a more comprehensive level of understanding of the underlying processes and their interrelations. During an 18-month period, cheap to change and process understanding, coupled with implementation of the nine principles, provided some significant benefits:

- The overhead head count was reduced by 20 percent. The reduction resulted from decreasing product cycle time from 16 days to 9 days for one family of products and from 3 days to 1 day for another.

- Over the next 18 months, additional improvements were made in reducing overhead. Under its divisionalized structure, each cumulative doubling of output reduced overhead by 25 percent. With the group manufacturing structure, each doubling yielded a 40 percent reduction.

The cycle time reductions in this case were not a result of developing a major new effort in concurrent engineering or cycle time management; they came about because HP manufacturing looked at the system and its processes and applied the principles. The integration of functional groups into a manufacturing team allows an organization to eliminate many of the start-up problems that consume time and delay product introductions. Edmondson and Wheelwright quote an HP senior plant manager:

> As you eliminate the number of questions that have to be asked every time a problem arises, you find you can get a solution must faster than you did before. Increasingly you can focus your attention on the high leverage opportunities rather than chasing an endless stream of low-level problems.

IBM

Specialization can be a blessing or a curse; today it is a curse for most organizations. Cross-disciplinary and cross-functional expertise is a rare commodity. The current emphasis on teams is even limited to the lack of breadth of knowledge and experience. A team consisting of 10 people from different disciplines is quite different from a team of 10 people who have cross-disciplinary and cross-functional knowledge and experience. One of the factors that allows Japanese teams to be more successful is that they do not make the distinctions between science and engineering that are made in U.S. industry.

Gomory, in "From the 'Ladder of Science' to the Product Development Cycle" [9], relates the history of IBM's turnaround of the Proprinter, a dot martrix printer for the IBM PC. When IBM announced its PC in 1983, a compatible printer from IBM would have cost $5,500, or about twice the cost of the PC. IBM decided to supply a Japanese-made printer. It then organized a team and gave it the charter to redesign the Proprinter. It would not be allowed the luxury of a multiyear cycle for completion. At the time, the Proprinter consisted of about 150 different parts and many different motions and actions. The group finally decided that the printer could be limited to about 60 parts if some parts performed several functions. It went through the usual redesign process by reconceptualizing, eliminating, and changing. Manufacturing was part of the team, and it concluded that the assembly could be automated if the proper design considerations were included. The result was a printer with 62 parts that printed faster than competitive products and provided added features. The equipment was reliable because it had fewer parts, fewer adjustments, and fewer assembly errors. The cycle time was reduced by 50 percent from comparable IBM projects in the past. Even though the printer had been designed for automated assembly, it was so simple that it could easily be assembled manually.

IBM is well equipped with sophisticated simulation facilities that may not be available to all businesses, but Gomory suggests that lack of such facilities "should not obscure how shortening cycle times is mostly a matter of managerial common sense." Management has a major responsibility in the up-front work relative to technologies, markets, and the business. It has the responsibility for clearing the track of obstacles that divert the teams effort from the basic objective. But management's role must go beyond removing obstacles. Contrary to some thinking, my experience indicates that continuous and timely involvement and review with the highest levels of integrity are an absolute necessity.

Modular construction and the use of standard components is certainly not a new invention, but for some unknown reasons, management practices, like the wheel, are reinvented. Gomory notes that going back to those concepts reduces cycle time. He cites an IBM case in which one group, working on a series of new display panels, reduced its development time by 5 months over previous development cycles for the same line of products by using standard components. However, managers must also make certain that they are not just receiving a modified version of an old concept because that is the easiest way to meet the new time schedule. Giving an old set of drawings some new thought with the intention of using as much of the old design as possible may not be the long-term solution. A competitor may be in the process of obsoleting that technology. It is important that needs of that new design be carefully considered and documented.

Another example from IBM involves the integration of groups, which is one of the major barriers to improving cycle time. Elmes and Wilemon, in "A Field Study of Intergroup Integration in Technology-Based Organizations" [10], present the findings from field interviews with 86 department heads and project managers in mature technology-based organizations. They found that:

- 38 percent rated the quality of group relations the same or worse.

- 53 percent rated the quality of group relations better or much better than in the past.

- 56 percent said their organizations were operating below the levels of capability, which resulted in delays, loss of efficiency, high costs, and too much politicking.

The study also revealed such contributing factors as lack of understanding of different goals and priorities, lack of a clear organizational structure, frequent changes in goals, strategy, and management, not working on the right things or working on the right things inefficiently, and the proverbial poor communications. All those factors affect the three elements of cycle time. Before cycle time can be accelerated in situations that display those characteristics, major changes must occur.

Intergroup conflict between the so-called corporate research laboratories and the divisional laboratories often exists. The conflict can be characterized by the relationship that exists to greater or lesser degrees between the corporate central research laboratories or advanced engineering groups and the R&D functions in the operational groups: Corporate related groups do not direct sufficient effort to technologies related to the operational units, and the operational units lack the competence to recognize the value of the technologies being developed in the central laboratory. Gomory, in "Moving IBM's

Technology from Research to Development" [11], relates how joint programs of central laboratories and the operating units foster technology transfer and provide a means for introducing next-generation technologies.

Although attempts were made to integrate the corporate and the operational research effort, progress was not evident. Certain ad hoc arrangements existed, but no formalized effort was directed at establishing the gateways for collaboration. The usual attitudes prevailed. Research felt that if the operational groups did not want to accept the technology, so be it. Put it on the shelf and forget about it. In those days, according to Gomory, where research was investing corporate resources did matter greatly to the rest of the company. In 1970, research explicitly took the responsibility for technology transfer, with the intention of bridging the gulf between research and operations. Two trips to Japan convinced Gomory and his colleagues that IBM needed to improve the way it transfers technology. That was the beginning of the first interdivisional joint program, called the Advanced Silicon Technology Laboratory (ASTL), as an umbrella organization to put into practice the next generation of silicon technology. ASTL included IBM research and two component laboratories under a single management and a single technical plan. The plan was eventually approved with a commitment from IBM research and the operating divisions. Gomory states that: "We rarely transfer technology anymore in IBM research. Instead we develop it jointly with our colleagues in the product divisions." Although no figures are given, IBM claims that introduction of technologies has been accelerated and product development cycles shortened. In 1989, IBM had 19 joint programs of IBM research and the IBM line businesses.

The third IBM example involves results related to the laptop computer. Cook, in "IBM Laptop: From Design to Product in 13 Months" [12], provides the following statistics of the achievable by direct numerical control of machining processes produced directly from the 3-D design data:

- The normal 2-year cycle reduced to 13 months.

- Mold production time reduced from the normal 16 to 20 weeks down to 10 weeks.

- Electronic data interchange (EDI) allowed engineers in two locations to work on the design concurrently.

- Computer-aided engineering (CAE) was used extensively.

- IBM declined soft tooling because of confidence in the team and the technology used.

Motorola

It is always interesting to speculate why it takes executives so long to see the light of day. They are intelligent, proactive, independent, and supposedly profit-oriented, yet they and their organizations get caught up in minutiae and refuse to look in their own backyards. Motorola provides a good example. The company blamed Japan for so many of its problems that the United States imposed duties as high as 106 percent on companies accused of dumping cellular telephones [13]. At an officers' meeting in 1979, a senior sales executive said: "Our quality level really stinks." George Fisher, the CEO, notes that: "We went through a denial process, trying to shout the poor guys down." The issue was rationalized: Motorola's quality was good, it was not good enough. Such situations are repeated in every organization whenever an attempt to rationalize a failure to meet expectations is made. However, Motorola decided to do something about it.

Motorola launched a major quality effort in 1980, and by 1986 it had reached a level of 6000 to 10,000 defects per million parts. That represented an improvement from 10 percent on the low side to 1000 percent on the high side. Motorola's 1992 target is to meet the six sigma requirements (3.4 defects per million). One can argue the merits of such objectives, but Motorola believes that some production lines in Japanese factories are operating at the six-sigma level. Motorola claims that their quality program yields improvements of $400 million to $700 million per year. The waste figure affects the cycle time of many activities. It accounts for the direct loss of time in each of the business functions. Fisher estimates that the quantifiable quality-related costs are about 10 percent of a typical organization's sales.

It was generally assumed that a new product would require 3 to 5 years from drawing board to the marketplace. In 1985, the engineers set a target of building a new paging device in 18 months. They called the project Operation Bandit [14]. Its guiding principle was to use as much off-the-shelf technology, regardless of its source, as possible and to search for the sources of excellence in and outside the industry. The objective was to have the highest quality and the fastest delivery factory. The customer information system was linked with the factory; an order could be communicated directly from the field. The process that previously had taken 6 weeks now required 2 hours. This example relates not only to cost but also to cycle time. Fisher admits that significant progress has been made in reducing cycle time on selected projects, many of them by 50 percent. But he points out that breakthrough products have not yielded the same results.

Motorola introduced the six steps to sigma as an approach to dealing with the nonmanufacturing related operations: what is the product, who is the customer, who are the suppliers, what is the map for fulfilling the mission, eliminate the non-value-adding activities, and establish a means for measuring performance and drive for continuous improvement. It is interesting that although the six points have been related to quality, they are really the foundations of managing. They go beyond their application to quality. Motorola's effort focused on quality, but the six steps yielded significant benefits in cycle time. As an example, the time to file a patent generally took 6 months and often took a year or more; it was reduced in some cases to a few weeks. Finances reduced the closing process from 2 weeks to 4 days.

NEC Corp.

Product development in Japan is a market-driven process. L. H. Young, in "Product Development in Japan: Evolution vs. Revolution" [15], notes that new products are shaped by competition, market philosophy, and complex supplier and distributor relationships. Those relationships have been well known for many years. Tatsuo Ohbora, a principal of McKinsey & Co.'s Tokyo office, states that over 90 percent of development work in Japan is incremental. What sets Japan apart from the rest of the world is that this improvement process is continuous. Attempting to tailor products for market niches is described as product churning. The approach focuses not on a market niche but on market niches. Japanese companies focus a great deal of attention on having their design engineers learn just what is important to the customer through direct contact with customers. NEC Corp. claims to provide customers with whatever product they want. Young includes a case study from NEC Corp. that demonstrates many aspects of the Japanese approach.

In 1987, NEC decided to introduce a new model of its portable telephone. A 10-year technology forecast was the first requirement. The prediction was that digital devices would replace analog devices by the end of 1990. By 1993, a single instrument would add such functions as paging and cordless service through satellite transmission. A detailed worldwide market study followed and was supplemented with other issues related to government regulations, and so on. Historical competitive data were collected, and the conclusion was reached that size was important to the customers. Each competitive phone was analyzed for size and plotted against the date of introduction. They chose a size of 270 cm^3, which was just below the current lowest size available, and projected a 150-cm^3 unit for introduction in 1991–1992. They froze the concept in July 1988 after exploring it with its subsidiaries in key markets. By December 1988, the preliminary specifications were converted to specifications from which a production model could be built. The model was completed in March 1989 and went into full production by December 1989. Even though additional customer input was received during the development process, it was not incorporated until a year later when a new model was introduced. According to Young, the product is a market leader and sold 300,000 units worldwide during the first year.

Northern Telecom

In 1985, Northern Telecom's PBX production facility in Santa Clara, California began a major renewal program in manufacturing. Grant, Krishnan, Shani, and Baer, in "Appropriate Manufacturing Technology: A Strategic Approach" [16], relate some of the Telecom history. The company did not jump on the computer-integrated manufacturing (CIM) or the advanced manufacturing technology (AMT) bandwagons but instead went back to the basics: It focused attention on the purposes, objectives, and strategies of manufacturing as related to corporate objectives; it considered the available resources and infrastructure in relation to the activities; and it recognized the needs for education, cross-functional integration and so on. The going-back-to-the-fundamentals approach yielded the following benefits from 1985 to 1990:

- Engineering changes could be incorporated within 48 h from time of notification
- Throughput time was reduced from 2 to 4 weeks to 2 to 5 days
- New product introduction time was reduced from 24 to 13 months
- Gross inventory declined by 54 percent
- Circuit pack yield was increased by 66 percent
- Work in progress was reduced by 68 percent

Target

The need to manage cycle time is not limited to the high-tech industries. Rouland, in "Target: Partnerships in Progress" [17], relates the case of the discounter Target and its programs with vendors for a quick response. VF Corp., which includes Bassett Walker, Lee, and Wrangler, is cited as an example. Target is attempting to build a partnership that would engage in joint up-front planning, share sales and inventory information, optimize the logistics, and reduce total network inventory by cycling orders more frequently. The program is driven by cycle time reduction (quick is better), responsive innovation, and open communication. VF's objectives include reducing cycle time by 40 percent, inventory by 30 percent, and cost by 20 percent. Target has now established similar objectives. This is an attempt to instill some real-time considerations into merchandising. Thus far VF has a 70 percent reduction in lead time with Wrangler, all three companies are shipping orders 98 to 99 percent complete, and production time at Bassett-Walker has been reduced from 13 weeks to 6 weeks. A purchase order generated at Target to Wrangler results in a 2-day turnaround time for receipt of order to shipping. The entire replenishment through Target's chosen carriers takes 4 to 7 days.

Thornton Equipment Company

Process and the need to consider the organization from a process perspective has been a theme throughout this book. Blaxill and Hout, in "The Fallacy of the Overhead Quick Fix" [18], raise the issues of escalating overhead costs in manufacturing as organizations automate or introduce new administrative capabilities. They state that reducing overhead is not about cost; it is about process. Without engaging in a major discussion of overhead costs, the authors do provide a good case study that relates to managing cycle time.

Thornton Equipment Company is a large Midwest manufacturer of specialty equipment: axles, engine and transmission shafts, and so on. Because of margin pressures, Thornton decided to attack the overhead costs. Programs included outsourcing a few models from each component group, which would save $3 million. The company reduced its overhead staff by about 10 percent, but 6 months into the program, the costs began to increase. With fewer staff, scheduling and maintenance suffered. To solve the problems, the company rehired some of the old staff. Management disregarded any actions to improve processes and postponed upgrading. Outsourcing required tracking orders, shipments, and billings. The added suppliers required more indirect staff, and those actions also resulted in excess capacity. The company closed five of its 15 plants. The net result: higher overhead. The cost in time of outsourcing is seldom considered. Furthermore, most organizations fail to take into consideration all the additional activities that detract from the supposed benefits of outsourcing. Process cannot be ignored. Not only does it affect cycle time; more important, it affects business performance. Blaxill and Hout introduce what they call robust process. It includes all processes such as design, manufacturing, and logistics. It includes all the business processes associated with the tripartite organization model as presented in Chap. 2.

Robust processes yield better cycle time, higher quality, and improved performance. Toyota averages 90 percent uptime in one forging operation; the best U.S. average hovers around 50 percent. Toyota's overhead, which averages less than one person per $1 million in sales, is the lowest in the industry; U.S. competitors average five people per $1 million in sales.

After 2 years, Thornton's top management came to the conclusion that, as a result of the outsourcing program, quality was suffering, costs were rising, and sales were declining. The conclusion was that outsourcing core components was counterproductive to the company's long-term goals. A team began rethinking the manufacturing approach and brought back many core components for in-house production. Thornton had pioneered many of the core processes, and to give them up to outsiders without the necessary competence was a major mistake. Management evidently looked only at the figures and gave little consideration to the levels of competence required for producing quality parts and assemblies.

The team found that competitors were achieving low costs by managing their processes internally. Thornton began a process change effort that involved determining the sources of waste, performing cross-functional training, using operators as sources of data and process control, upgrading and introducing new manufacturing technologies for better flexibility, and redesigning the plants for a more optimal work flow. Some outsourcing of marginal low-volume products and low-value-adding parts and assemblies continued. A long-time vice president of R&D (24 years and a well-respected industry visionary) was asked to resign. His ideas about the role of process in adding value were not congruent with those of management.

Within 5 years, the results were quite different from those of the outsourcing program. Thornton realized $18 million in sustainable annual overhead savings without compromising volume or product diversity. Also:

- Total overhead employment dropped 30 percent over 2 1/2 years
- Supplier base shrank by a factor of 3
- Unit cost dropped by an average of 18 percent
- Design engineering time for new parts decreased by an average of 34 percent
- Engineering overhead to direct engineering was reduced by 26 percent
- Product development time was reduced by 40 percent

Toshiba

Consensus decision processes or collaboration? The discussion of consensus versus collaboration is a never ending process without any definitive conclusion. Rehfeld, in "What Working for a Japanese Company Taught Me" [19], provides some clues to different approaches to decision making in Japan and the United States. Rehfeld cites the case of the need for compatibility with IBM. Toshiba was in the process of reentering the computer market after some disappointing experiences. Compatibility with other equipment, and especially IBM, was a major consideration. Not even such companies as Texas Instruments, Digital Equipment, Wang, and Hewlett-Packard were compatible. The question of Toshiba's new entrance into the computer market involved a decision relative to compatibility. Rehfeld believed that Toshiba America had to be compatible. Toshiba's Japan engineering in the Tokyo factory and Tokyo marketing thought otherwise. A study team was organized by McKinsey to survey distributors, dealers, and users relative to the need for compatibility with IBM to gain consensus among the Japanese and obtain their buy-in. The response was a definite yes for compatibility. As a result of the study and follow-up discussions, Toshiba Japan not only bought in on the need for compatibility but totally committed itself to the development of the first full-featured IBM-compatible laptop computer. How much time can an organization spend attempting to reach 100 percent consensus? There is no doubt that consensus is essential and that the Japanese follow the principle; at least, that is what the literature implies. Perhaps this is a Japanese blind spot that at some future time may be modified. U.S. industry may at times be at the other end of the spectrum: little consensus; just start to do something. Rehfeld suggests that although it may be appropriate for the Japanese to seek 100 percent consensus, perhaps a more collaborative approach in which 20 percent of the effort achieves 80 percent of the buy-in might be more appropriate. The challenge is to have the right 80 percent. As a practitioner, I have found that waiting for 100 percent consensus on any decision consumes inordinate amounts of time and adds little value. The S-curve applies. A point of diminishing returns is reached, and it takes competent managers with a track record to know when to stop using limited resources for marginal returns.

Velcro

Velcro [20] produces those hook-and-loop devices that fasten children's sneakers, the fabric ceilings in automobiles, the headrest cloths on airplanes, and many other things that need fastening. Velcro's business involves the automotive, textile, and medical industries, but its principal activity is automotive. In August 1985, Velcro management received a call from General Motors (GM) that it had 90 days to set up a program on total quality control in its U.S. plant. A quick trip to GM revealed that Velcro was providing a fine product and there were no problems with delivery. GM's concern was too much inspection after production. GM's dictum was build the quality into the product. Scrap amounted from 5 to 8 percent, depending upon the product.

A crash course was given to 500 employees on the importance of quality, problem-solving techniques, statistical process control, and the use of information from detailed records. At the end of 90 days, a plan was presented to GM. It detailed where Velcro was in its total quality control program and where it was going. GM people listened and said they would be back in 6 months for another review. However, the dumpsters continued to be filled with scrap and waste. Waste was reduced in 1987 by 60 percent from the 1986 levels, and by 1988, there was another reduction of 45 percent from 1987. The company skimmed the cream off the top. Quality control reduced its staff from 23 down to 12, but the basic manufacturing process remained unchanged.

Attempts were made to increase the speed of the machines, but the scrap rates increased to unacceptable levels. *Levels of improvement are limited by the process.* Additional benefits require changing the process as well as the mentality of the organization. Manufacturing the Velcro product not only requires producing the raw material but involves dyeing to specific color standards, coating various materials, and eventually slitting and cutting into specific widths and lengths. Velcro production employees knew that days could go by before defects were uncovered. Thousands of yards of product could be scrapped. When an organization produces 8 percent scrap, machines are running longer than required, valuable resources are being wasted, time is being allocated for non-value-adding activity, and the process needs some attention. Somewhere along the growth track, Velcro management lost sight of available technology. That is not an uncommon situation. At the time of the GM request, measurement and instrumentation and control systems available for nondestructively testing every inch of product that went through the system. At that time, there was not nor is there now any reason why a machine should produce scrap or out of specification product for several days. The Velcro example illustrates two points in relation to cycle time: (1) the cost of additional time related to the three elements of cycle time and (2) the need to recognize the limits of improvement possible if the same basic manufacturing system remains in place. How could Velcro management become so satisfied that it failed to make major process improvements? It was a single-source supplier without competition until its patents expired in 1978. Taiwan and South Korea got into the act, and prices, which they controlled at one time, plummeted by 40 per-

cent in 2 years. Competition changed the environment, but it was predictable.

Use of Special Tools

Tools appear to be the means by which organizations attempt to improve
performance. The interests in reducing cost, improving quality, and reducing cycle time are interrelated. Tools and techniques can play a major
role in compressing the design time. The net effect depends on how the
tools are used and how the released time from using them is directed.

Computers in the auto industry

There is no doubt that computer-aided design (CAD) can provide some
benefits, but the benefits depend on how the system is used. If it is
merely a replacement for the drafting board, the benefits may be questionable; CAD may be just another high-tech toy. Fitzgerald, in
"Compressing the Design Cycle" [21], relates some examples from the
auto industry. She notes that the use of computers and CAD has not
shortened the overall design process for the auto industry, although individual design segments take less time. Converting to CAD or any of
the other computer-related tools requires education not only in the new
tools but in changing the culture. If computers make it possible to
shorten design time, the hours saved must be directed toward some
new productive activity. Fitzgerald relates the comments of Gary
Whitaker, a CAD-CAM support specialist at Ford. Whitaker designed
chassis parts before and after CAD arrived on the scene in 1979.
Although the new technology reduced the time to design the rear sus-
pension from 3 months to 6 weeks, CAD users spent the rest of the time
exploring design options. No one in management was concerned about
design cycle time.

Prior to CAD, the design of a component as simple as a windshield
wiper required careful three-dimensional analysis of the relations between the wiper and the windshield. With current software programs,
designers now take the same amount of time to analyze 40 or 50 different possibilities. Richard Johnson from Chrysler believes that exploring the alternatives allows engineers to create high-quality
optimized designs. Although various computer-aided systems provide
significant benefits, the manner in which they are used cannot be ignored. Is it really necessary to develop 40 or 50 different possible designs of a windshield wiper? Computer-aided systems are not a
panacea. Designers need to think, and the computer does not provide
much assistance if the creative thought is lacking.

Whistler Corp., Bolt Beranek and Newman Inc., and Codex

An article in *Electronic Business* [22] relates the predicament of three companies that faced dramatic changes in their business environments: Whistler Corp., Bolt Beranek and Newman Inc. (BBN), and Codex Corp. Whistler Corp. is a consumer electronics vendor that discovered its Spectrum radar detector about to be released to production sometimes malfunctioned in extremes of temperature. The condition was uncovered about a month before the company was ready to ship the product. It took another 10 months to fix it. Whistler was already competing against the Asian companies that outperformed it on price, time to market, and knowledge of customer needs. At the same time, radar-detector prices were dropping at the rate of 25 percent per year. To remain in the business, Whistler reduced the work force from 600 to 200 and cut R&D spending to 1 or 2 percent of sales. Whistler did not have a history of managing product development. There was very little oversight. The company hired a consultant to study the company's product development process. The essence of the findings was that product development was too slow for the type of business and that it occurred in starts and stops. Technology was viewed only as a means to an end: there was no technology strategy. The study also recommended that the company develop a platform of core technologies that could be used across product lines and adopt quality, function, deployment (QFD) as a means of determining how many functions to put into a new product and aid in making the trade-offs. Whistler began a program called WINS, which was directed at product development that included a phased review process, use of core teams, and adoption of QFD. The benefits that accrued were:

- The first product developed under WINS was accomplished in 11 months, 10 months less than the previous average time.

- The first phase review was held 10 months before completion. In the past, it would have occurred at the latter part of the development cycle.

- Quality was improved. Five years ago, 9 percent of Whistler products failed and were returned by the customers. The current number is.04 percent.

BBN depended on the government for a major share of its business. It developed a cost-plus mentality that was not compatible with industrial management practice. It did not understand the significance of operating in a market-driven environment. As the company attempted to expand its computer-related hardware, software, and communications technology, it became apparent that an acceptable product development infrastructure did not exist. Having recognized that the cost-plus system was inadequate in the commercial marketplace, the company developed a program designated Edge. Edge was basically a new way to approach product development with phased reviews, core teams, and so on. It emphasized thinking and planning before embarking on physical development. The new approach involved changing the culture of BBN. The process uncovered one project that was almost a year behind schedule and no longer met current market requirements. It is difficult to understand how an organization like BBN, even though it was heavily involved in government contracts, could function without an organized product development process. This BBN example clearly illustrates the negative effect on the three elements of cycle time. How many worker-hours beginning at the top of the pyramid will be needed to change the culture? The hours will not add value; they will consume time and detract from the output until the product development process is in place and operating in a new culture.

Codex, a division of Motorola, adopted a product and cycle time excellence (PACE) program to improve its product development capability. The management concluded that six sigma and PACE were two sides of the same coin. Improvements in quality affected improvements in product cycle time, which in turn improved quality. Codex first implemented PACE in late 1988 with phase reviews and core teams in several development projects. It expanded to include all major projects in 1989 and all projects in 1990. In early 1991, the company had 72 development projects, all supervised by core teams. Codex claims to have reduced its average product development time by 46 percent from 140 weeks 2 years ago to an early 1991 average of 74 weeks.

Japan's experience with QFD

Quality function deployment (QFD) is another tool that, although related to quality, affects product development cycle time. Burrows, in "In Search of the Perfect Product" [23], says that QFD puts the drawing board into the customer's hands. It is a means of helping organizations make the trade-offs between what the customer thinks is needed and what the company can economically provide. It is something that has been done and continues to be done, but usually in a more erratic and nonstructured way that allows preconceived notions and biases to dominate the decision process. QFD is a matrix that allows product decision makers to determine what is important to the customer. That matrix includes customer preferences on one axis and customer priorities on the other. The objective is to search for the important relationships. Like CAD, the system must be used intelligently. Although the structured approach provides benefits, the results depend on the complexity of the matrix. Burrows notes that many companies have ended up using a 100 by 100 matrix, which means filling in 10,000 cells. Such an approach is a waste of time. It does not add value. Like various scheduling methods such as PERT and CPM, the exercise is not the end result. The purpose is to determine the critical and important relationships. The bells and whistles may not be necessary: They may only cause additional field problems with no observable benefit to the user. QFD can improve the new product introduction success rate because, as Michael Anthony states, it replaces the "voice of the engineer" for the "voice of the customer." Although QFD puts a great deal of emphasis on the customer, astute marketers and technology collaborators are needed to separate the chaff from the wheat. The customer may not always be right. Translating prevailing thinking into facts and reality involves a risk. The majority opinion or perception of reality may or may not be projecting into the future. The buggy makers did not participate in the auto industry. The auto industry did not participate in the airline industry. Too often, the people who are in the forefront of marketing and technology, those who really know, are ignored.

According to Fortuna, in "Beyond Quality: Taking SPC Upstream" [24], QFD was first used by Mitsubishi's Kobe Shipyard in 1972. It provided a means of resolving many of the factors that extend product development cycle time: differences among functional groups, long development cycles, and personnel changes. QFD provided visual information that could be easily interpreted by the downstream participants. Many Japanese companies use QFD. Toyota reports design cycle time reductions of 30 percent and start-up cost reductions of 61 percent

1. Engineering changes reduced by 30 to 50 percent
2. Design cycle time shortened by 30 to 50 percent
3. Start-up costs reduced by 20 to 60 percent
4. Warranty claims reduced by as much as 50 percent
5. Additional user benefits:
   - Improved customer satisfaction and market share
   - Better designs and performance
   - Transfer of engineering know-how to next generation products

Figure 12.2 Savings and benefits achieved by Japanese companies by using QFD. This does not imply that QFD was the sole contributor to the benefits.

during 7 years that included four start-ups. Aisin Warner claims a 50 percent reduction in engineering changes and design time. Komatsu MEC, a heavy-equipment manufacturer, introduced 11 new products within a 2-1/2 year development cycle. Tokyo Juki Kogyo Co. used QFD to help redesign its line of sewing machines. It became number one in the sewing machine market in a long-term declining market. Figure 12.2 illustrates the savings achieved by a sample of Japanese companies by using QFD. Like any tool, QFD must take into consideration the particular situation to which it is being applied and the constraints placed on the system. It is a means of identifying conflicting design requirements. It is not an end in itself; it is not a single-issue management tool.

References

1. K. B. Clark and T. Fujimoto, "Lead Time in Automobile Product Development Explaining the Japanese Advantage," *J. of Eng. and Techno. Manage.*, vol. 6, no 1, 1989, pp. 25–58.
2. A. Taylor III, "Why Toyota Keeps Getting Better and Better," *Fortune*, Nov. 19, 1990, pp. 66–79.
3. I. Nonaka and M. Kenney, "Towards a New Theory of Innovation Management: A Case Study Comparing Canon, Inc. and Apple Computer, Inc." *J. of Eng. and Technology Manage.*, vol. 8, no. 1, 1991, pp. 67–83.
4. M. Moritz, *The Little Kingdom,* Morrow, New York, 1984.
5. J. Teresko, "Corning's Rebirth," *Industry Week,* Jan. 7, 1991, pp. 44–47.
6. N. Tichy and R. Charan, "Speed, Simplicity, Self-Confidence: An Interview with Jack Welch," *Harvard Bus. Rev.*, vol. 5, September-October, 1989, pp. 112–120.
7. S. P. Sherman, "The Mind of Jack Welch," *Fortune,* Mar. 27, 1989, pp. 39–50.
8. H. E. Edmondson and S. C. Wheelwright, "Outstanding Manufacturing in the Coming Decade," *California Manage. Rev.*, vol. 31, no. 4, Summer 1989, pp. 70–90.
9. R. E. Gomory, "From the `Ladder of Science' to the Product Development Cycle," *Harvard Bus. Rev.*, November-December 1989, pp. 99–105.
10. M. Elmes and D. Wilemon, "A Field Study of Intergroup Integration in Technology-Based Organizations," *J. of Eng. and Techno. Manage.*, vol. 7, no's. 3&4, 1991, pp. 229–250.
11. R. E. Gomory, "Moving IBM's Technology from Research to Development," *Research/Technology Manage.*, vol. 32, no. 6, 1989, pp. 27–32.
12. B. M. Cook, "IBM Laptop: From Design to Product in 13 Months," *Industry Week,* Aug. 5, 1991, p. 57.
13. C. Morgello, "George Fisher of Motorola: The Quest for Quality," *Institutional Investor,* August 1991, pp. 45–46.
14. R. Henkoff, "What Motorola Learns from Japan," *Fortune,* Apr. 24, 1989, pp. 157–268.
15. L. H. Young, "Product Development in Japan: Evolution vs. Revolution," *Electronic Business,* June 17, 1991, pp. 75–77.
16. R. M. Grant, R. Krishnan, A. B. Shani, and R. Baer, "Appropriate Manufacturing Technology: A Strategic Approach," *Sloan Manage. Rev.*, vol. 33, no. 1, Fall 1991, pp. 43–54.
17. R. C. Rouland, "Target: Partnerships in Progress," *Discount Merchandiser,* January 1991, pp. 37–39.
18. M. Blaxill and T. M. Hout, "The Fallacy of the Overhead Quick Fix," *Harvard Bus. Rev.*, July-August 1991, pp. 93–101.
19. J. E. Rehfeld, "What Working with a Japanese Company Taught Me," *Harvard Bus. Rev.*, November-December 1990, pp. 167–176.

20. K. T. Krantz, "How Velcro Got Hooked on Quality," *Harvard Bus. Rev.,* no. 5, September-October 1989, pp. 34–43.
21. K. Fitzgerald, "Compressing the Design Cycle," *IEEE Spectrum,* October 1987, pp. 39–42.
22. "Three Companies That Have Seen the Light of Product Development," *Electronic Business,* June 17, 1991, pp. 65–68.
23. P. Burrows, "In Search of the Perfect Product," *Electronic Business,* June 17, 1991, pp. 70–74.
24. R. M. Fortuna, "Beyond Quality: Taking SPC Upstream," *Quality Progress,* June 1988, pp. 23–28.

Epilogue:
The Time-Driven
Enterprise

Using time as the force that drives the enterprise is a most powerful way to gain business advantage. Organizations that emphasize time surpass their competitors, but the benefits depend on how management perceives time. Focusing on time does not mean speeding up activities and working harder at doing the same things. The time-driven organization goes beyond using simplistic concepts, premises, facts, theories, opinions, biases and prejudices, and relationships in the decision-making process. It elevates the substantive issues to a higher plain. That new level involves knowing the limitations of all the elements and rationalizing the known with the unknown, the controllable with the uncontrollable, and the predictable with the unpredictable. That state of knowledge requires integrity. It mandates a more comprehensive perspective of reality. It is rooted in the fundamentals of management practice and not a quick fix or this year's major-emphasis program.

The time-driven organization exhibits the characteristics necessary to cope with delineating strategies, managing transitions, surpassing the competition, meeting the quality expectations, exploiting and leveraging technologies, expanding into new or global markets, integrating people and functions into a cohesive system, involving the whole organization, and the many other factors that affect organizational performance. It operates as a system that uses its infrastructure and resources to achieve certain specified objectives. It knows its limitations, its strengths, and its weaknesses. The time-driven organization neither acts like the hare nor the tortoise. It acts sometimes like the hare, sometimes like the tortoise, and generally at some point between the two extremes. Slow and steady may be applicable sometimes; taking the giant leap may be applicable at other times.

Considering time as the force behind a business may resurrect the ghosts of the efficiency experts and the often derogatorily referred to principles of Taylorism. The stopwatch and the clipboard are not the tools of the time-driven organization. Thinking, doing the up-front work, synthesizing information, using the results of introspective business analysis, providing leadership, and participating and contributing are some of the tools. Being time-driven does not imply trying to pack more work into an 8-h day. It does include using the least amount of time and meeting requirements at the right time and within an acceptable time period. It involves working on the right things, the things that add value. It focuses on the time required to move something through the system. The time-driven organization is not obsessed with time, it respects time and treats time as a valuable resource that cannot be replaced.

The time-driven organization focuses on effectiveness, efficiency, and the economic use of resources. Not all the activities that make for a successful business will be quantified by the accountant's pencil. Many will be qualitative and equally important; some will be more important. The short-term and the long-term needs must come into balance. The time-driven organization focuses attention beyond the technology-related activities. It goes beyond time to market, simultaneous engineering, and time compression. It includes the total system. It rationalizes the inconsistencies and inadequacies of all the components of cycle time. It eliminates the barriers and balances the diverse requirements of the various constituencies like the elements of a Calder mobile.

The process of choosing to operate as a time-driven enterprise begins at the top. Any business consists of inputs, a process, and outputs. The inputs consist of the resources and the infrastructure that will be acted on through some process to provide a specific output. Inputs can be transformed into acceptable outputs only if the process is based on adequate content, a system for measuring the various parameters that affect performance, a feedback and feedforward capability, and a control system that keeps the process operating according to some predetermined criteria. It involves more than going through a set of procedures. It requires thinking; it requires disciplined competence; it requires a full understanding of the doable. At times, the process may tend to drift out of control, but that is why feedback and feedforward information is essential to minimize the impact of those interruptions on system performance.

An organization that chooses to become time-driven does not need to begin a major series of studies by acquiring data, information, and knowledge about itself and its competitors. It does not take months to acquire the essential information. Know the fundamentals, get 80 percent of what is needed in 20 percent of the time, and add a good measure of human initiative and creativity. There is no need to copy the Japanese. Most organizations have the basic talent; they only need to listen. It begins at the top with the CEO and the various executive committees. Educating the participants is essential, but education by example should not be ignored. Moving to a time-driven organization is a matter of learning and doing and learning and continuing to learn. Managers in a time-driven organization:

- View the enterprise as an integral system.
- Emphasize system performance results, not functional performance.
- Pursue activities that add value.
- Leverage resources through an effective business infrastructure.
- Manage the business process—substance, outputs, and feedback.
- Foster organizational vitality, resourcefulness, and creativity.
- Create a challenging environment—freedom with business discipline.
- Raise expectations at all levels and in all disciplines.
- Improve effectiveness, efficiency, and economic use of resources.
- Influence, guide, and coach the participants.
- Understand the internal and external business dynamics.
- Infuse proactive leading; reject reactive managing.
- Build an organization with a sense of urgency.
- Challenge the organization constructively.
- Live in the real world and reject egocentric rationalization.
- Balance change and stability.
- Rock the boat, but know the limits.
- Demand competence at all levels.
- Eliminate the gimmicks—the fads—the single-issue management.

The time-driven organization demands dedicated and competent proactive leaders. They must firmly set both feet in the business: one foot cannot be in the office while the other is turning toward some non-value-adding activity.

The time-driven organization cannot focus on single issues. It cannot focus on *time* as a single issue. First and foremost, it is guided by the fundamentals of management practice. Experimentation in the managing process is not something to be avoided: Like experimentation in the laboratory, it is an essential requirement for improving business performance. But experimentation in the dynamics of managing must be grounded in fundamentals of experimental design. It must include more than "Let's try it and see what happens." From that perspective, meeting the requirements for exploiting cycle time for technology or business management is a result achieved by practicing the fundamentals rather than conceptualizing a new program.

Index

ABOUT THE AUTHOR

Gus Gaynor brings more than 35 years of technical and management expertise to the process of integrating technology into business operations. A 24-year career at 3M, Minnesota Mining and Manufacturing, provided him with extensive executive-level experience in managing technology in a global context. He is principal of G. H. Gaynor and Associates, a technology management consulting firm. Mr. Gaynor is a recent Fulbright scholar in the management of technology. He is an active memeber of the Academy of Management, the Engineering Management Society of the Institute of Electrical and Electronic Engineers, the Society of Manufacturing Engineers, and others. He serves on the industrial advisory boards of several universities and professional journals. Mr. Gaynor resides in Minneapolis, Minnesota.